<u>FORWARD</u>

I0462313

DEDICATED TO THE ONE FOREVER
GLORIOUS INFINITELY BENEFICENT, ALL
BE-LOVING, MERCIFUL ORIGINATOR AND
LIFE BESTOWED OF OUR MAN SPECIES
AND ALL UNIVERSAL BEINGS

CHAPTERS

"PRAISE THY EVER BOUNTIFUL CREATOR

WHO CREATED MAN FROM A CLOT,

AND TAUGHT HIM BY THE PEN,

WHAT HE KNOWETH NOT."

SURA 1R "THE CLOT" "AL ALAQ"

PREAMBLE

This book is decades in the making, nay a lifetime, albeit thousands of pages must now be condensed into my present script yet, however I recreate anew, always the same all-compelling inner voice inspires - reveals and reverberates the same message, and in actuality, elected mission.

Never more timely to my 21st centurion generations of man! We know more about the dark side of the moon than we know of Islam - and its universal living communities of moral voracity and peace towards Muslim-kind brethren and non-Muslim humanity, despite the devil jihadist terrorist's et al and their barbaric ISIS evil perpetrators of crimes against humanity.

May God's all-impelling, omniscient and gracious overwhelming love and truth's revealing wisdom, guide this non-conforming ignorant and humblest servant, in the noblest blest service of my brethren man: aye, ever in pure awe... Celebrate, praises and gratitude ad infinitum to the glorious magnanimity of humanity's forever beneficent creator cherisher, for this unsolicited, inexplicable ... Incredible marvel of life being!

CHAPTER ONE
"IDENTITY"

Humanity's incessant, obsessive quest for his or her self-identity has awe-propelled, perplexed, plagued, and confounded relentlessly - the human species, since earth's birth! Man is born with his implacable, insatiable all-compelling irrepressible quest to know who he or she is; the why, wherefore, and from whence he or she comes, and to where or whither he or she goes! And the non-self created human enterprise yet mesmerizes the where-withal of the undeniable "who's who" originator/creator!

Islam, more than any other religion, creed, theosophy, philosophy or ideology, reveals man's royal "Who's Who" inherent divinity! It defines man's God innate divine nature, "Al Fitrah"/God. Man's indivisible and inextricable unity and ineffable oneness, with the Uncreatable, unique, and perpetual eternal one "Al Wahad", whom we share eternity, by virtue of our God-endowed nature. Al Qur'an reveals the heart of Islam. (Sura 30:30)

"Man is created in God's own nature... Al Fitrah" and endowed with God's self-same attributes and qualities... Of love, com-passion, beauty, peace, liberty... "All of my beautiful names", reveals God to his man emissary and trustee... Of universal nature and all living beings. Mankind, possessing his creator's God-potential, for the perfection of his man-species embodies no "original sin"... But man is "saved", by virtue of his own God-infused, inherent Nature - "Al Fitrah".

"Know thyself," wisely said the pagan Greeks, only they didn't explain or elaborate on who and what, we of humanity are, in terminology of identity. Islam reveals man's identifying nature of being. "To know thyself, you must know God ... And to know God, ye must know thyself" for man is forged and indissolubly linked to his Originator, Sanctifying Creator.

Islam never suffered the disparate duality of western religion, secular division and separation of life from religio-spirituality within the commonality of everyday existential living. Islam is without church, temple, or mosque as institutionalized. Man, individually constitutes his own priest... In direct access to God, sans intercessor or hierarchal authority other than God within. God is neither extrinsic nor extraneous, but ever present, intrinsic to man's interior core and fibre. "The kingdom of God is within you" reconfirms Islam's revelatory edict that "God dwells in the heart of his beloved man believer."

Islam institutionally reverts. Converts, and consists of a living vital, moral, veracity of family and community on an international status quo, comprising a vast diverse ethnic, racial plurality of global nations... In a world community "Al Dar" presently exceeding 2 billion of the populace.

Since missionaries or proselytizing ... Is strictly forbidden by Islam... Yet incomprehensible to Christendom ... That Islam is the world's fastest growing religion! Its appeal speaks to the commonplace man, who intuits God readily available within the truth of his own being.

We, in the west, remain grievously incognizant, with all our sophist educational criterion and electronic, computerized, communicative technological "information age" - abysmal awareness of Islamic civilization, culture, and science evolution, expanding over 15 centuries; whilst darkling, medieval Europe suffered the abyss of ignorance and superstitions of witchcraft, demonic forces, and vampires. Islam, through God's revelation, actually "authored", proscribed and impacted the scientific knowledge, researches and advancement for immutable truth for mankind, which made possible the basis and foundation of our modern scientific technology. God's revelatory commandment for Islamic humanity, and all ignorant Homo sapiens... "To search beyond earth's parameter and the celestial heavens, for enlightenment of truth ... This be thy holy quest and incumbent duty... From the cradle to the grave" and "for my knowledgeable man creature "Al In'Saan" "the first thing that God created was reason!"

Less than a century after the Holy Prophet of Islam, Muslim geologists had measured the circumference of the Earth, as a round sphere, dispelling medieval Europe's false notion as flat, and in 8 A.D., the apogee of our planet's sun, moon and sea tides were measured with mathematical precision.

While Arabic Muslim scientists had revered and preserved Greek theoretical sciences, it was Islamic scientists' transposition of Greco-scientific theories into Islam's mathematical genius,that radicalized, advanced, and uniquely discovered and created new scientific mathematical disciplines and axioms. The new equations of axiomatic database enabled Islam's astrophysicists to ascertain mathematical exactitude, the apogee of planet earth, to sun, moon and mercury, in 8-9 A.D. before pre-Renaissance medieval Europe had even begun to discover or assimilate Islam's scientific transmissions to the West.

It is uniquely Arab Islamic science's esoteric spiritual abstract eye, that produced the Arab genius of mathematical sciences that identified and formulated the intrinsic essences of qualitative properties of universal nature, in perfect concordance with extrinsic numeric extrapolations, and providing the exigent key, wherewith exact concrete measurements of spatial, linear distances and light years transmitted from solar galaxies could be approximated. That religion, in tandem with science, achieved one of innumerable scientific advancements is nothing short of amazement and consternation to our western science secularist, Godless chasm, yet unholy disparity between science and religion! Religion, to boot, the cataclysmic contributory to scientific enlightenment!

Reiteratively, the spiritual abstract Islamic genius, that discerns through the "soul's eye"... "Ruah Al Ayan" can decipher between the fabric of temporal mutable materiality and the dimension of spiritual immutable reality, "Al Haqq", only thereby, pierce into cosmic nature, cognize and recognize the light of spiritual reality and perpetuity even herein, man's present earth tense. The universal natural is supernatural here and now man rides astride his earthly capsule, implanted both on the earth and celestial bound, with heaven's seal and royal crown!

For us Westerners, de-spiritualized materialists of contemporaneous science, we must attest, that man's light of spiritual eye is singularly vital, and exigent to mankind's uncompromising advancement on abstruse blind sight and ego pseudo-expertise!

We, of American humanity, are in a crises of soular identity: spiritual depravity and degradation of our once virile proud Judeo-Christian moral veracity of sustaining values and character ethics, have plummeted to corrosive, materialistic, complicit conformity, pervasive now in common citizenry, and the professional elite academic and scientific authoritative consensus of so-called "expertise". The new religion of modern man is "egolatry": worship of ego supplants any/all vestiges of earnest aspirations for true enlightenment of knowledge and the in sighting meaning and perspicacious eye of immutable truth. Over-laden and overcome with technocratic ever-accruing factual database going nowhere in a dead end spiral of nonsensical!

The true "Scientia" science sought by our forebears' sacrosanct commitment to Eternal Truth, are stricken down in our new science of soulless, Godless, irreconcilable fixation, or some would call transgression. Our drug-infested medical therapies have abysmally failed to cure America's endemic mental-neurological emotional illnesses, anxieties disorders and morose depressions; degenerative diseases, conspicuously cancer accelerates daily, hourly, extinguishing precious lives of our American populace, withal our medics' expertise, or rather "ego expertise" failing to discover the underlying causes thereof; and non-committal to nature's inviolable truths, re: the amazing human organism! Of our human spiritual, physical ecological immune balance and equilibrium ascribable necessarily to universal nature's genius creator, conceiver of the human brain/ body marvel!

Our western sciences have defunct ineffectual "secular" modes of soulless, Godless mechanist hapless tragic death warrants for countless, however inadvertent or unawares! Western man is our planet's most endangered species!

We are verily indeed in a soular crisis of identity! We suffer from soul paralysis. The soul of man is marginalised; nay denied, in western medicine … as some vague "psychic" phenomenon, the province of religion, and western science's express collision with religion, and secularist science schism with religion and the God of religion. In direct contrast, Islamic medicine, "Unani": recognizes the whole soul, mind embodied human organism, vital to prognostic therapy, healing and cure; and the "perfect physician" genius God creator's supernal spiritual empowering remedial cures, contingent on peerless optimal health and longevity! While "meditation" and other mental exercises render calmness more peaceful attunement with man's higher spiritual propensities: positive healing and real cure can only fructify with direct ineffable oneness in love, through prayerful repartee and joyful intimacy of unity in the One unique God/Love Creator. For love alone uniquely cures! Islam's science of medicine attests and has proven over centuries, man's unifying alignment in God, ignites love's overpowering cures and ensures illness and disease preventative arsenal!

We can, applying Islamic medicine's "Unani" therapy, realize unprecedented salving healing and cures vis a vis America's deadly Cancer and degenerative disease endemic. Precious, untimely death sentences can virtually be overturned. If we Americans restore our hereto-fore faith and spiritual moral veracity and core religious values, committing once again, our spiritual fervor and faith to God, our nation's creator, founder of American humanity's extolled attributes of God-imbued, inherited character ethics, love, brotherhood and peace.

Lament and deplore, I do, for our national American community's once formidable social cultural mores of neighbor conscionable collaborative solidarity and collective morality. Our once peace abiding streets, now besieged and rampaged with rampant crimes, drugs trafficking and addiction, wanton senseless violence, rape and gun fun-toting psychopaths! There are no safe places of refuge, for our beloved children's innocence. Even in our schools!

And our once proud family citadel's sanctuary, stricken down with sexual abuses and assaults, even that heinous crime of all, incest committed against our most vulnerable and innocent! When once, we were a "nation under God"... Endowed by our creator, and vouchsafed sanctifying trust and holy virtues consciousness!

To lift the shroud of common knowledge relative to Islam: Islam is not a "new" or yet "another" among the world religions; but from the Abrahamic, Semitic lineage of Abraham, Moses, Israel and Ishmael, prophetic chain through Isaiah, and culminating with Jesus, "the Christ messiah", to Muhammad, prophesied by Jesus, as "the Christ counselor of Truth", who would fulfill his messianic mission, and implement his teachings and Commandments. Thus God revealed to the holy prophet of Islam, he, Muhammad was "the seal of the prophets"... No other prophet would follow the Hebrew-Christian-Islamic revelations, thereafter.

While Jesus is revered and venerated as "God's Messiah" to Islam and all mankind, Islam reconfirms the mosaic revelation to worship the one God... To take no "associates" or iconic gods comparable to the one sovereign creator... Nor Jesus, deific as "triune" in Christian litany, for God is infinite, above and beyond human mortality... Neither flesh nor limited in being. Jesus, is defined in Islamic revelation, as "the compassionate breath" and purest spirit of God, lighting mankind". And "the Christians of the golden scripture are closest to you, O Muslims". Repeatedly over and again, Islamic revelation repudiates Christendom for violating Jesus' own "First Commandment": "Thou shalt worship the Lord thy God first, with all thy heart, mind and soul." The "Shahada" worship of the one omniscient supreme God, reverberated in Muslim prayer, re-stores and reaffirms the Christ Messiah commandment albeit, the God of the Torah defined as a "jealous God"; Islam redefines God as the "beloving cherisher creator" infinite love's creator; who requites man's uncompromising, indivisible oneness of love! One illustrious Ibn Arabi declared, "Love/God is the cause of all love. Were it not for love, God would not be worshiped!"

God yearns that man search and discover knowledge of his Uncreatable man Originator Cherisher. "I was a hidden treasure. I created man, in order to be known!" Therefore "seek knowledge

14

beyond Earth's orbit... To the zodiacal heavens... And cosmic nature and ye, will find, O man... I am the same infinite beloving, cherisher creator, who dwells in your heart!"

By virtue of Islam's heart of revelation, that man is imbued with God's own innate nature, "Al Fitrah"; God designated man, to govern, in God's stead, as "royal trustee", emissary and holy steward of universal nature and being. Al Qur'an reveals that God commanded all the angels to bow down before his "man creation as ruler in my stead"... All prostrated, excepting Qur'anic Satan, "Shatan", the jealous and prideful, fallen angel.

One European scholar of Islam likened God's reflection of himself in man, with these analogous words. "The stars would not know their glory, were they not reflected in the waters below!"

We will endeavour to divulge some, however minuscule, of the vast corpus of Islamic scientific discoveries and radical trans-formations and advancements that made possible the basis of our own western technological sciences possible. Despite the exclusionary void of thousands of Islamic scientists, and voluminous scientific sources, never translated into English. As Islamic sciences began to disperse and disseminate into medieval Europe, igniting the renaissance, Ibn Sana's (Avicenna) "Canon of Medicine" replaced the old wives' tales and fallacious quackery, becoming Europe's bible of medicine.

Islamic sciences, inspired by Al Qur'anic revelatory principles and axioms of scientific discovery never suffered the West's fatal nexus of science and religion; for Islam never exercised as "religion" per se, but the underlying truth of all universal nature and Immutable Reality, "Al Haqq", Islamic science encompassing the science and nature of both mutable and immutable reality.

Revelatory ad-lib, were the inviolable intelligible laws and governing flawless wisdom of universal nature... Unalterable invisible living spiritual essences, operating in perfect concordance from the minute atom to the vast, magnificent celestial solar galaxial blazing star clusters and constellations in the ever creating influx of dark energy's rebirth of solar nebulae and new galaxies. The inner, unseen world reality "Ayan" precedes the material embodied reality. Thus, earthly beauty never dies... a "rose" conceived in heavenly pre-existent imagery, never dies! The supreme artist who inspires our own puny artistry ... Relegates eternally peerless beauty... "Al Jamil", one of God's surnames.

As with man's attendance of angelic forces, whose flawless intelligible omnipresence "services" man, every waking, sleeping moment, revitalizing his or her immune system, magical voices communicating, uppermost "TLC", tender loving care... For the likes of you and me! Our vital organs and cells... "Hot wired" through the amazing human brain/body complex... Unified to

16

an obvious Beloving, Caring Omniscient Creator Cherisher!

"I am closer to you, O man, than your jugular vein!"
"My abode reigns in the heart of man."
Love, infinite does indeed circumnavigate man's every pulsating fiber!

Only now is Western enlightenment beginning to trickle into our academia and science compendium... That Islamic civilization, over one millennium, had successfully founded the world's first egalitarian society - where race, ethnicity and diverse religions... Christian churches, Jewish synagogues, and mosques flourished side by side, in communal calm and brotherhood, "Umma"; in Islam's recognition and societal, political, juridical proscription and mandatory edict, that true equality thrives legitimate-when all ethnic, racial plurality of humanity flourish as "one human family". Reiterating the Holy Prophet's edict... "We are not Muslims or non-Muslims, whatsoever the ethnic or racial complex... But humanity is one and the same family." Islam's non-racist identity of all humankind, living in harmony, unification of community and social equilibrium... We in the West can better emulate, at this post "civil rights" still discordant and discriminatory injustice.

While Isaac Newton was toying with gravity observations of falling fruit trees apples, Islamic geo-physicists had discovered electron nuclear magnetic fields of the earth, gravitational pulls of

sun, moon and inter-planetary orbital ellipses...
The biosphere and stratosphere, terrain, and ocean
tides' magnetic protective insularities centuries
before western sciences had discovered and
harnessed electro-magnetic components into
electricity. From 8 A.D. to 9 A.D., electric
generators lit the streets of Cordova, Spain.

El Burundi's genius, as with other numerous
famous Islamic physicists, who yet remain
unknown to our western science enclaves, had pre-
Renaissance, discovered spatial light magnetic
curvature, and the inner esoteric invisible light "Al
Nour" God creating hidden forces charging visible
nebulae and cosmic matter, we now call "invisible"
or "dark matter" and "dark holes", in scientific
lexicon.

The relevancy of cosmic time and space was
pronounced centuries before Einstein's "Theory of
Relativity"; only Einstein's momentous discovery
entailed time and space relative and confined to
visible matter, excluding the invisible spiritual
substance energies and essences ... Islamic
astrophysics identified as "Al Batin" physics, in
perfect complement with the visible in flux
materiality, "Al Thahir". There constitutes the
immutable constant axiom and perennial scientific
principle and premise of the cosmos intelligential
forces and agents of "Al Hakim"/God... Flawlessly
executed throughout the universal cosmic
operation. From the unseen molecular atoms,
integral to all visible cosmic living forms,
"particles" and entities, tacitly "conversing" their

18

cellular internet communicative reshaping and recreating… consummating... "As God wills."

As the reader will surmise, Islam never suffered the science versus religion disparate, divisive dilemma of Western science. The marvelous cosmos, and the marvel of the non-self-created man, indubitably signaled unceasing awe and wonderment at the Supreme Architect and flawless engineering execution, to be celebrated and praised... "Subhan Nallah" praises the cosmos originator in endless, awe-rapture of incredulous splendor, and unceasing wonder! The down-to-earth, concrete Muslim scientists would poetize in celebratory lyrics, God's perfect glorious marvels manifested in their research discoveries and deliberations, intermittently praising the God Originator inimitable Conceiver/Creator in their science mundane texts.

We will reiterate throughout this script, Islam's science of "Tawhid" - universal nature's inviolable law of indivisible unity of relative mutable materiality and absolute spiritual immutable reality; so desperately, sorely needed in our Western science's spiritual dearth and death nemesis of in transient mechanistic malaise! Science's de-spiritualized theorem consensus afflicts every avenue of contemporaneous life, from failing to elucidate the spiritual veracity of our universe, to curing our soulless, Godless diseases endemic crises! The ego-techno God of science demagoguery has ensnared our heretofore light of intellectual creativity and truth's incentives towards the underlying, immutable veracity of our cosmos and our liaison connection, as the man creation! To what noble or ignoble purpose... This amazing life's gift and marvel of being? "I have not created this marvel of life for you, as a joke... O man!" Saith mankind's conceptual maker-benefactor... "But with express purpose and noblest goals" reiteratively.... "Ye, O man, are my greatest miracle... Why look ye for anything less?"

Einstein confessed, after many distinguished years of his momentous scientific discoveries, his sole regret was in not proving God's existence and wholly operational omnipresence in the universe. He rejected scientific validity of quantum mechanics and its depiction of reality in terms of probabilities, as introduced by the German physicist Max Born. "Some physicists, among them myself, cannot accept the view that events in nature are analogous to a game of chance... God does not play dice in the universe."

With the increasing transmission of Islamic scientific knowledge and enlightenment into Europe, and igniting the renaissance, Islam, with its non-institution, non-hierarchal religious freedom, and vested authority in God alone, stupefied the medieval church, and posed a direct threat to western church papacy control and authority. Competitive, discriminatory hostility ensued, clearly curtailing and robbing Western commonplace humanity of the freer spiritual veracity and religious ready accessibility to God within, and the metaphysics of Islamic physics, contributory to Europe's fledgling scientists, as a whole spiritual and physical science ecology of universal nature and man's correlation therein. Had the whole integral science of spiritual and physical counter-balance been initiated to European scientists, already suffering religious control, suppression, and persecution from church autocracy, it is reasonably probable, that Western science would have encompassed the whole integrated metaphysical and physical Islamic scientific axioms, instead of our Western mechanistic paradigm abyss, and de-spiritualized technocrat demagoguery!

George Bernard Shaw predicted the "Islamization" of Europe, as he espied Islam's irresistible appeal to the common man, sans church or institutional intercessor or man-imposed intermediary, other than God as sole authority within man's spiritual being. Shaw foresaw Islam as Christendom's greatest challenge and stimulus for reinvigorating and revitalizing Christianity within the context of Islam, which he deemed posed no conflict, with Islam's acceptance and veneration of Jesus Christ as the Messiah. And peradventure, Islam was indeed the Paraclete which would implement Christ's mission.

The rule of law, (we modernists speak of so redundantly) originally founded by Moses' God revealed summons, and consummated by the Holy Prophet of Islam, 1500 years ago, before the Magna Carta ... The "Sunnah... Or "Shariya": civil, political and judicial laws, proclaimed and implemented true equality to all ethnic, racial, religious and cultural diversity, of commonplace humanity. Islam liberated all slaves and abolished slavery (inscribed on our own United States judicial monuments); granted custodians for orphans, ensuring their rightful inheritance, granted independence to women, securing women's rights under the law, individual, pre-marital inheritance and property ownership, holding express title to their land and properties, retaining their maiden names, instead of losing their identity in their spouse's family name (millennia before "woman's lib") and equal custody of their children, in tandem with their offspring's fathers, rights granted western women, only decades ago. This book will delineate in more comprehensive historic detail, in the chapter on "womankind", Islamic women acclaimed and distinguished as poets, philosophers, mystics, scholars, etc., honored by their male peers; and Islam producing the world's first woman physician doctor in Muslim Spain.

The corpus of laws, under the "Sunnah" is too vast and voluminous to include herein this short space. There were commercial moral stringent requisites for merchants and business entrepreneur's uncompromising honesty, fair and just economic dealings with public consumers, in all aspects, weight and cost exactitude, fresh produce and untarnished products, and prohibition and illegal food price gauging. Any and all property rights must be derived by mutual honest consent... And exploitation of property values or rightful acquisition... Severely condemned and rectified to the victim. Brotherhood, "Umma", solidifies all human encounters, relative to all life's activities, and all relationships, beginning with family sanctity, relatives, neighbors, friends, and lastly strangers, Muslims and non-Muslims. Brotherhood warrants unity, unification and peace with all the "family of man". "Zakat" is foremost the act of charitable giving to one's brethren... An involuntary "tax", incumbent on all Muslims, to render monetary charity, according to his means, a fair share portion of his financial property or income, to provide for the poor, indigent, sick and incapacitated. If a Muslim is unable to pay Zakat, or fiducial funds, he is then obliged to give charity, by acts of moral assistance and conscionable giving of self; feed the hungry, assist the frail, elderly and handicapped, by loving kindness. One adulteress was forgiven, for her charity giving a dog, dying of thirst, using her worn garment, to draw water from a well for the animal. All animal's plight, whether domestic or wildlife, were entrusted for safe-keeping stewardship, and the prohibitions gravely adhered to; as pitting one animal against another for sport cruelty or any harmful infliction committed against nature's innocent creatures or under-lings. Severe penalties and incarceration swiftly followed. One man was imprisoned for abusing a cat, by his village, and damned in heaven as well. All good acts of express kindness earned redemption here in the present as well as afterlife. The Holy Prophet of Islam directed all Muslims to greet each other with salutations of peace, "Salam" and wear an open cheerful face and optimistic manners, persona and faith, as an act of charity.

God revealed through Al Qur'an, that the Holy Prophet was the most exemplary of moral excellence, for mankind to emulate. Noteworthy to mention here, when Western, including the Parisian newspapers caricature of the Holy Prophet incense Muslims worldwide, in the name of "free speech"... Westerners truly divulged their ignoramus consensus of Islam, to poke fun and make mockery of the Holy Prophet of Islam, is regarded, not only an abomination, but betrayal and moral violation of "free speech", which carries grave responsibilities of verbal degradation and denunciation of "dignity" afforded any human being, let alone the venerated Prophet. While violence never justifies pursuant any irrational ignorant act of our misguided fellow journalists, words, insulting, demeaning and contemptible, do veritably assail and assault, in lieu of knives, guns or weaponry. Of course, must I reiterate the Western divisive conception of religion and God versus Islamic reality that religion infuses every segment of everyday life? And God is all-pervasive in Islamic international society! We must make allowances and forgiveness for Westerners' gross abysmal-pervasive ignorance!

The list for moral conscionable directives is endless... To enumerate further... As afore reiterated, Islam is intrinsically indivisible from every day life... Islamic law incorporates mankind's highest moral veracity of human social, civil, cultural imbedded prerequisites... Islamic societies are beholden to the world's utmost God-oriented segment of humankind. Contrasted to our own Western secularity of life, divisive from religio-spiritual mores and moral tenacious criteria, and greater freedom to commit highway robbery, illicit business practices, corporate monopolistic greed, and all the other endless easy means to deceive and defraud our fellow Americans. But, let us recall, there was once an America of religious ethics pervading conscientious Christian entrepreneurs - the Carnegies, Fords, Roosevelts, et cetera, and an American nation, sanctified under God.

"Shariya", or "Shariya law", distorted today, by new alleged Muslim power-demonic degenerates... "Stoning"... "Beheading" or the like is nowhere in Qur'anic or Shariya law. "Shariya" literally translates as "the righteous path of justice, compassion, mercy, equality, liberty, and equity in all human enterprise and interaction..." including equitable economy... and jurisprudence. This "righteous mean", "Shariya", revealed in Al Qur'an, and instituted by the Holy Prophet, consecrated and guaranteed for "all the family of mankind"... However desecrated or flagrantly violated by pseudo falsifiers and evil perverters of real Islam and Islamic law.

"Uqrah!" or "Read!" Erupted the First Commandment to the God elected prophet of Islam, 609 A.D., that forever changed the entire course of world history!! "Uqrah!"... ("Read!") Gabriel, heaven's archangel. Twice commanded...

Towering over Muhammad, in failing cover of light, his pulsating breast grasped by the all empowering angelic host, in near awe-breathless suspense. "Uqrah!" ("Read!") Came yet again, that commanding utterance to El Muhammad, who retorted in his quivering voice... That he was not of the learned and literate ones, but among the plain, humble and meek of humankind.

Then "Recite... And praise thy most gracious bountiful Creator, who raised and created man from a clot, and taught him, by the pen, what he knoweth not!" This first Sura of the holy Qur'an is revealed to Muhammad... "Al Aqal"... "The clot."

Muhammad, still dazzled, overwhelmed, as in a feverish state of tumultuous emotions, returned to his homestead and beloved wife, his trusted companion and confidant... Disclosed his ordeal. Who is he, an ignorant, unlearned insignificant of mankind, chosen by the Singular One God Creator for whatever miraculous divine purpose and mission? Khadija, knowing her spouse's pureness of heart, and unrivaled reputation for integrity, moral veracity and trustworthiness, so surnamed "Al Amin" by his fellow Meccans, assured Muhammad, that indeed, he had been elected and chosen by God, as God's holy messenger for God's great, as yet unknown mission, and purpose. Muhammad now recalled a surprise encounter at the age of five, while astride a caravan... Halted by a holy Syrian saintly Christian who laid his hands on the boy child's head, and declared him, to be one day, "the Christ

paraclete ... The Christ counselor of Truth",
prophesied by Jesus Christ, who would follow
him, confirm and fulfill his Christ mission on
Earth. Muhammad's astounding disclosure, and his
visitation of God, revealed purpose for the Holy
Prophet, convinced Khadija and she was
admittedly, Islam's first disciple.

Having tragically lost his father, even before
his birth, and his mother, at the fragile age of six,
Muhammad was left an orphan of the noble tribe
of Quraysh, raised by his faithful uncle, Abu
Baker. His oftentimes loneliness drove him within
his spiritual propensity, searching for an
immutable reality and God creator, intuitive
sovereign presence within his purest heart of soul.
The youth was wont to seek spiritual solace and
solitude in one remote cave on Mount Hira...
Where he meditated endless hours, forgetful of
self, family or community, ever in search of truth
and a living God creator. Grappling with the
Meccans' worship of multiple idol Gods, and
Mecca's corrupt commercial exploitation of idols
and idol worship. Conspicuously, the Kaaba, now
the center and trafficking for idols worship, he
knew from historic annals, had been Abraham's
makeshift temple, celebrating God, after
Abraham's son, Ishmael was spared his life.
Muhammad, through Ishmael, was Abraham's
direct descendant... And long hoped, the Kaaba
would be reconverted to a holy monotheist Semite
religion of one eternal living God. Oddly enough,
Muhammad foresaw his own prophetic realization,
before his God summoned mission had even

begun!

That revelation adjunct the cave in Hira, overlooking Mecca... Destined to irreversibly alter and evolve world history. Heralding Islam's posthumous monumental sciences achievements and enlightenment that would ultimately eradicate Europe's darkest ages, and usher in the foundation of modern technological sciences!

The resounding Islamic mandate, the quest for "El Ilm", the knowledge and science of enlightenment; incumbent on all Muslims, became a rallying obsession; the search for the universality of truth, "Al Haqq". The Holy Prophet of Islam said, "The pursuit of knowledge is a divine commandment for every Muslim. To spend more time in learning is better than time in praying... It is better to impart knowledge of wisdom, one hour in the night, than to pray the whole night. One hour's contemplation on the work of the creator, is better than seventy years of prayer!" And the Holy Prophet revealed, "The ink of the scholar is more holy than the blood of the martyr!" And so inspired and expedited Islam's famous worldwide scientific expeditions, probing and researching all diverse social, cultural, religious, and political vistas of all extant human societies... From the Arabian peninsula, through North Africa, stretching beyond the sub-continent of India, into the Asian Caucasus and china! It was, verily God, who initiated mankind's research into universal, immutable truth and reality! ("Al Haqq"/God) For man's self-discovery was pivotal, in discovering his own God inherited, imbued nature. Revealed repeatedly the Holy Prophet... "He who knoweth his own self knoweth God!"

Not by the sword, as fictitious erroneous western historic sources belie: but by 'proselytizing enlightenment' and scientific truths... The Muslim emigrating ventures pursued, their biologists, geologists, physicists, scholars scientific and religious concierge, engaged the indigenous populace, assimilated and synthesized their vast diverse societal cultures' chaff from the kernel eye of correlating principles of truth, thereof, recognizing the manifold forms, and faces of manifesting truth, latent within every human soul and segment of mankind. Conspicuously revealed, per Islamic tenets... To be discovered in all climes and places. "We have not created a people, nation or community, to whom we have sent holy messengers, savants, admonishers and prophets". Truth is universally inherent in all mankind. Buddha, Confucius, and unknown prophets, and God inspired, honored emissaries lauded, among some 200 plus, are revealed by Al Qur'an.

Wherever the Muslim entourage and new immigrants traveled, they respected religions and faiths of all kind; preserved the Jewish synagogues and Christian churches, once again recognizing the "Umma" egalitarian brotherhood of all "the family of man". Islam pursued intellectual conquest of enlightenment, absorbing, synthesizing, civilizing and transforming status quo, and stagnant civil societies anew, soliciting moral codes of veracity and heightened criteria of social justice and a new egalitarian solidarity of community, multiplicity of ethnic and racial unity. Enlisting the Jewish and Christian intellectual gnosis and illustrious minds, together in scholastic coordination, produced signal excerpts of Islamic scientific corpus into Arabic, Hebrew, Latin, and neo-Greco translations for Europe's succeeding Renaissance.

Conquest, not by the sword - as depicted by fallacious western historical sources - but by intellectual enlightenment and scientific engendered truths, disseminated heretofore-vast African, Asiatic and Anglo European fledgling nations. Islam's inestimable contributions to scientific knowledge and universal enlightenment, only now, to our present day, finally acknowledged, albeit in diminutive and scanty assessment.

The Muslim mogul, or "Mughal" military incursions and conquest of the Indian subcontinent, of Delhi, Agra Hindustan, the Deccan, Kabul and Kashmir provinces, yet remained respectful of the existing Hindu and Brahman cultures, and conferred praiseworthy honor to Buddha's God-inspired revelations and wisdom. Only idolatry was deemed abhorrent fallacy and desecration vis the one ever living, self-subsisting eternal God creator. "Truth has come, and falsehood vanishes, nil nullified and naught of Immutable Reality... Al Haqq"/God.

It was also here, that Islamic architecture

evolved to its zenith, at Agra with the glorious world famous "Taj Mahal" built to celebrate Shah Jihan's beloved Mumtaz. Taj Mahal, the 7th Wonder of the World, this most regal architectural marvel incorporates Islamic "Jinnah" "heaven's" depiction of celestial vaults, ever transcending - with lush, intoxicating, over-laden gardens and pure shimmering flowing waters, sweet bliss. And peace, peerless, directly imaging and reflecting Al Qur'anic magnifying revelation of heaven's portals and future paradise, has never since been imagined, or translated into any other human art form. My western peers complain of women's "inequality" injustice by Muslim men! When in stark reality, what western, European, or American man would ever erect such a memorial celebration of womankind, for you, or me, sisters, and offspring of western culture? The so unlikely truth, for my sisters to ingest or realize, is that Muslim womankind are idealized, elevated, and elated far above mankind, deemed their moral superiors, yea, "superior" to men, however disbelieving to our popular culture and deplorable gross ignorance.

Wherever and whenever the Muslims ventured, immigrated, integrated, conquered or colonized, the new Islamic promulgated civilizations, inaugurated, disseminated and practiced their preeminent governance of moral humanitarian core precepts of unequivocal equality for all diversity of ethnic, racial, social, cultural, religious humanity, within Islam's new egalitarian order "Umma" brotherhood for all "the family of mankind"; which community solidarity, incorporated the obligatory equitable economic justice for the impoverished, destitute and marginalised underlings of human being Diaspora. When Muslims founded their new "Umma" social order of brotherhood in the Indian sub-continent, they were astonished to witness India's "untouchables"; and soon reversed their heinous status to humanitarian equality - a significant historical fact, disclosed by Gandhi, in praise of Islam's egalitarian equality of brotherhood.

When the 2nd Khalif, Ibn al Khitab, entered Jerusalem, not a drop of blood was shed, his famous declaration historically recorded by Jews, Christians and non-Muslims alike, "We come in the name of the same God, of your Hebrew Torah, and Christian scriptures, confirmed by our Muslim Qur'an, in the holiest name of humanity's one beneficent creator". (Ibn Khitab's famous entrance into holy Jerusalem has only been recently known by the U.S. "Time-Life" series.) Ibn Khitab forbade his men to enter Christendom's holiest shrine, the Church of the Holy Sepulcher, in reverence to Jesus Christ, Islam's own accepted, venerated "Messiah", and lest his Muslim entourage convert the sacred shrine into a mosque; and bade his men, to pray on the outside hallowed grounds. It is note-worthy to add, that several centuries thereafter, Christians cognizant of Islam's veneration for the "golden book of scripture" and the historic sacrosanct Church of the Holy Sepulcher preferred Muslims to tend and care for the church, and only until the formation of the Jewish state of Israel, this shrine was maintained with utmost vigilance and protected by Muslim successive families.

The Khalif, kalifates, or governing amirs, were not monarchs or kings, as in our western connotation. For Islam identifies all humankind as "royal inheritors of God's own innate divine nature "Al Fitrah"Malaki yo maddine... God is the only... Sovereign king" as man reflects God's exultant attributes of unceasing beneficence, compassion and loving care and mercy, does he tread higher in virtuous esteem with God and amongst the common man. Wealth, power and fame, are of no avail, rank futile, in this temporal passing life-tense... Unless "spent... And used in God's apportioned limitless loving service to one's fellow man brethren". "Ye visited me, when sick. Ye fed me, provided and cared for me"... Is re-verbalized and re-verified by Muslim acts of benevolence and mercy, in the highest services for God's beloved sake!

"The only aristocracy is the nobility of the soul"... "Soul... Interchanges with the heart"... That is, "nobility or generosity of the human heart"... For God's magnanimity should mirror and personalize man's royal, God-derived attributes!

The Khaliph, or governing head of state, was recognized as the peoples' "trustee" representative, whose incumbent officiating duties were to safeguard the judicial, social, and economic justice and equitable equality of all the peoples, Muslim, or non-Muslim; to ensure and expedite education and medical services, freely provided, to all economic classes, regardless of property or material capacity. Islam, in actuality, has forever annulled "class systems", whatever his or her means, "equality" stems from all mankind's cherished esteem, in God's ever beloving, beneficent and merciful eyes, the Khalif was called to task, if he, or his judicial representatives failed to fulfill their solemn sanctioned moral obligations, so acutely uncompromisingly are the "people's power" vested and invested in themselves, and for the beneficence of their communities. Even within Muslim recent history, the late Amir Faisal maintained weekly appointments to personally oversee and hear any complaints, claims, or disputes of common citizenry. And in one of such claims by an army soldier, contesting Amir Faisal's land property as his own inheritance, after researching, and investigating, Khalif Faisal concurred with his plaintiff, and promptly relinquished his property, deeding title back to the Saudi, now rightful owner.

The Khalifate, administering the principal governing role in Islamic society, was oft versed and excelled in social, civic and cultural issues, education and scientific knowledge. They were foremost patrons of the arts and sciences, and held high court events and sessions with distinguished doctors, scientists, scholars, poets, and literary Muslim and non-Muslim creative artists; inviting the public at large, to attend lectures and discourses, honoring these renowned, and stimulating participation of the audience attendance. We westerners are vaguely familiar with the Baghdad Abbasid and El Rhahun Al Rashid's famous court, that solicited distinguished meta-physicists, physicists, astronomers, mathematicians, and other leading scientists, mystics poets, etc., who converged as emissaries for the advancement of knowledge and scientific truth. The entourage of scientific genius, such as Ibn Sina, Ibn Rushed, Ibn Khaldun, Ibn Razi, Ibn Arabi, Al Ghazali, accrue too vast numbers to mention in my short space... Plus the Khalifate heads of state - amirs, sultans, et cetera, who sponsored, fostered and promulgated Islamic art, medicine, jurisprudence, education and scientific efficacious contributions! A very few notable Khalkis, among too multiple to include herein, were sultan Abu Hassan and Abu 'Inan, themselves scholars, as well as Abu Battuta, geo-physicist from Tangier, Ibn Manzuq, historiographer, and countless others. The Khalifate founded some of Islam's most distinguished universities, Cairo's "Al Azhar", the world's oldest consecutive learning institution, and the Qarawigin University and Mosque, founded by Fatima, daughter of Muhammad Fihri, originally from Kairquan, Tunisia. His famous daughter, a scientist-scholar, inspired and expanded the religious and intellectual centre of Fez. Khalif Abu Ya'Qub in Marrakesh paid homage, honor, and patronage to the famous Ibn Rushd, as well as so many eminent philosopher-scientists. Islamic art and architecture came to its full illustrious splendor in the university mosque at Cordova, Spain. The famed circular domes and vaulted mosaic arches, emitting natural light and multicolored glass glistening mystical rhythms were premised on nature's "beehives" marvelous symmetry of construction; from which Europe thereafter copied and emulated, in erecting cathedrals with arches, sans supporting pillars. The ancients of Greek and Roman architecture had lacked the knowledge of construction, without their pillars' support. It was Islam's revelation of heaven's celestial ascent and infinite transcending irradiant light... "Al Nour"/God that created the new radical aesthetic transformation. Incidentally the blessed "bee", has its own chapter or Sura, in the Qur'an, "for man's great health benefit and instruction to emulate the bee's marvel of construct" henceforth, the circular dome and transporting arches heavenward dominated intrinsic Islamic architecture. Similarly, the Earth, and cosmic heavens moved with the elements in harmonious rhythm, peace and equilibrium.

The platonic ideal of the "philosopher ruler" was in some paltry way, characterized by the Muslim Khalifate. Only the Khalif was committed to God's inspired Qur'anic revelatory precepts, rigorous moral governance and humanitarian philanthropy implementation vis a vis pagan Greek secular methodology.

The Khalif was elected, not by political enclaves or parties, but directly by the collective "will" of the people themselves. Elected on the basis and grounds of his reputable, humane good deeds, his moral excellence and virtuous services towards his brethren man, throughout the towns, cities, and communities of the kingdom. Only thereafter, continuing his eminent governance, by the protracted will and consensus of the people, en masse. The Khalif could be easily overruled, and dislodged, by not exercising his moral committed ascendency, to the masses, or by the peoples' needed change for reform or solicited social advancement Islam and Islamic judicial law aptly provided for living changes, according to the times and generational options. Wealth was administered for the collective needs of the common people: the Khalif, albeit head of state, had no personal riches, other than provided for his and his family's basic living needs and sustenance. "Poverty is my pride," saith the Holy Prophet of Islam, who lived from day to day with only the bare necessities, although his coffers overflowed with his Bedouin tribes' and community's riches! The vast wealth, so accrued, was distributed to the poor and indigent, to the orphans, widows, and all in dire or unfortunate straights. The Holy Prophet's exemplary life and voluntary deprivation, was exemplary for the Khalifate to emulate and follow suit and, in fact, the second Khalif, Ibn Khaldun, was wont to disguise himself and visit the needy, sick, orphans and widows to discover any neglected act of moral obligation, or inhumanity.

Both pre-Islamic and post-Islamic trade and commerce had customarily established capitalism and free enterprise, for centuries on end provisions for the poor by "Zakat", as described before. A charitable taxation paid, according to one's monetary means, and fair share of profitable gain and income, as a primal moral obligation: state appropriation of private property, was denounced and rejected. Islamic judicial and political ideology of free capitalist enterprise, totally opposes communism: and the historic Muslim Russian Cassocks' fierce stance and battles against the Bolshevik Revolution of communism, is rarely portrayed in western historical annuls. Today, in China, the Yin-Yang and other Muslim-Chinese peoples are openly defying communist China and suffer endless persecution daily, from the Chinese communist oligarchs. Wherever state dictatorship or communist powers exist or wage tyranny and suppression, Muslims rebel and implant the seeds for revolution, and delivery for free democratic or egalitarian equality, and liberty.

After one thousand years plus, the peace-abiding Islamic civilization, from the Iranian and Arabian Peninsula, through Egypt and Tunisia, was ravaged by the barbaric hordes' onslaught of Genghis Khan. As world history records, the hordes pillaged and destroyed Islam's renowned libraries, schools, and university mosque complex centers, hospitals, science/technical citadels of astronomical observatories, and all edifices and instruments utilized to research continuously for scientific advancement.

Genghis Khan's descendants of the Turkish nomads, in one of history's great enigmatic twists of irony, later converted to Islam and rebuilt

Islam's prized science institutions of learning, and renewed scientific research, and judicial freedom for the common citizenry. Posthumously, the invading crusades confronted Islam's illustrious leader - the famous Saladin, "Salah'adin", "Defender of the Faith"; and astounded the medieval crusaders with Islamic civilization's far advanced social, cultural and judicial equanimity, science and medical technology, and moral pervasive religiosity, and robust, illustrious civilization! Moreover, Islam's respect and reverence for the gospel of "El Messiah" Jesus Christ, and the Islamic communities' welcoming posture and hand of peace, "Salam," only to be shattered by later militant crusaders' exploits of self-righteous spite, wanton destruction of Islam's holy citadels, and horrific massacres of Jews and Muslims alike - Jerusalem's indigenous inhabitants!

What is inexplicably amazing and wholly incomprehensible - nay, disconcerting, in assessing any civilization, by the conventional norms and conforming standards, is that despite the brutal tyrannical occupation of the Ottoman Turks, which followed their initial concordance and assimilation into Islamic civilization, but thereafter descended into suppression, subjugation, violence and chaos, against the innocent citizenry of the Turkish empire; and, following the First World War, in which the Middle Eastern Arabs fought alongside the Western allies against their Turkish enemies (well documented in Sir Lawrence's "Seven Pillars of Wisdom", yet, but once again, despite

"Lawrence of Arabia's" impassioned struggle and hope for the Arabians' nation-hood; we all know the infamous colonial powers' greed for foreign lands, properties and peoples' occupation and unjust regimes' unabated avarice for pure self-interest, viz common humanity. Yet, despite all the above adverse, historical realities perpetrated upon Islamic Arabia, 10 plus centuries of oligarchical, tyrannical suppression, subjugation and social, economic and political injustices, Islam prevails and the moral quintessence of its civilization yet subsists in Islam's uppermost "genetic DNA" spiritual character and temperament, inherent through generations. For when religion ceases to employ external props and doctrinal church litany and institutions, and when religion instead, congeals into man's internal spiritual efficacy, persona, and pragmatic reality, faith subsists within man's God-inspired holy intimacy repartee of mind and heart, within the beholding believer! For "the kingdom of God is within you"... And "God's dwelling lies within man's heart", re-echoes Islam. "Prayer is converse with God"... Saith the Holy Prophet of Islam... And all prayers' high design and purpose is ineffable, holy union... And love's oneness in God... "Mi'raj" the Prophet asserts that, "prayer is the union "mi'raj' with God or annihilation of ego in the Divine Essence and Presence!"

Unlike our western disparate division of religio-spiritual logistics, Islam's religion and religious faith creates a living affinity and exulting repartee of spiritual divine chemistry with God's effusing spiritual potency and character infusing persona reality!

Thus, "the Arab Spring" was not a whimsical passing phenomena, but representative of the never-dying, still-thriving Islamic living, viable, vital, and tenacious civilization, invested and inherent and intrinsic in the peoples themselves. Re-inspired by God's beneficent names, indelible within their conscience, and God bestowed attributes... Of "liberty"... "Al Thahir" named at the "Thahir" square and collective gathering of those rallying round new Arab Islamic social reforms for justice, "Al Ade'l" and equality. "All my beautiful names... Of liberty, justice, equality, beauty, wisdom, compassion, mercy etc. Have I, your God creator bestowed on my man creation"! All derived and inherited within man's God identifying innate nature, "Al Fitrah."

Predictably, there will be many more "Arab Springs" to come, as soon as the re-fired, revitalized Muslim youth and family communities rid themselves of their dictators, throw off their suppressive yoke of anarchist tyrants and terrorizing devil jihadists.

Uniquely, unlike other contemporary religions, Islam has infused and transfused into the common everyday culture, and moreover transformed the very psychic persona of Muslim humanity! "Secularity" has absolutely no credence, or eco-environs reality, and in the Arabic language, no word for secular or any semblance

41

exists.

The Islamic science of "Tawhid" - Islam's whole science of universal nature, unifying the esoteric spiritual and exoteric physical as one integral indivisible whole, has indeed passed into the Muslim common-place layman; in his mind-sight visage and heart-set unifying "Tawhid" inextricable unity of both the mystical spiritual immutable and material mutable realities. This should not come as incredible, inasmuch as we, in the west recognize the cultural infiltration of our own secularist, atheist Godless science, so alienated from religion, God and spirituality! Science does indeed bless with its underlying universal truths of enlightenment, or inflict its soulless Godless truth-less thesis upon everyday life and societal culture, so lamentably foreshadowed by our sciences!

God still configures as the "Who's Who" Godhead preeminence all pervading among global mankind, "Al I'lm" the all-wise, all knowing, uniquely self-subsisting uncreatable genius cause creator of our magnificence of cosmos and the likes of thee and me, who provides the universal subject matter for our fumbling misguided scientists. And the heart's intuitive truths that forever yearn for the immutable truths of our universe and mankind's correlating inter-connecting reality we can learn so meaningful enlightened knowledge and science of Truth and Immutable Reality, "Al Haqq"/God, from our Muslim brothers and sisters.

It is no wonder that Islam's international nations and communities remain the most God-oriented society of humanity on the face of the earth. Prayer, on his or her lips distills the early morn, with praises of remembrance ("Dik'kar") and celebration of the glorious one, magnanimous beneficent God Creator, cherisher and sustainer of life, and all universality. Prayerful "converse with God"... Be it silent or vocal echoes... Transcends mundane existence... From twilight to darkening dusk. Compound this daily, day-to-eve communal repartee with the lord God creator... Multiply your daily living alignment in unity with your "Ruah"... soul's... Soul... "Al Ruah" and environment over generations, creates a powerful heredity, and irreversible reality in the psychic spiritual condition in the dynamic virile soul of man! We can even synchronize a formula that could serve as maxim for the social scientist: environment or "e" times x generations or "g" equals = heredity "h". "e x g = h" the God- and man-oriented God daily tempo of Islamic society is incontestable and absolute reality. Muslim spirituality supersedes all our western notions of our own de-spiritualized strident secular polarization of our lamentable demoralized civilization.

Suffice to apprehend that the spiritual eye of insightful truth is the consummate fruit of spirituality. We will delve more deeply within these chapters, vis a vis mankind's higher evolution and evolvement, in exact ratio to man's spiritual light of intellect, and cognizance of the immutable truth of spiritual reality. Spiritual evolvement breeds consciousness of love for God/Good, man, compassion, peace, beauty, bliss and deathless certainty.

Mankind's prayerful dynamo impacting everyday monumental living, moreover augmented to customs and codes of God's remembrances, through salutations of "Peace", "Al Salam", upon greeting friends, acquaintances, even strangers, non-Muslims, remains indelible in Islamic culture and all humankind encounter. The recipient of God's gracious inspired greeting, replies with like brethren response... "Ou Alaika Salam" - "Peace abide with you, too." God's presence and wisdom is sought to attend any/all individual, family or social events, and all future endeavours, enterprises and undertakings invoke God's pervasive blessing and guidance, "Insha Allah"... And "Rah Matalan" in order to incorporate, even a fraction of God's al Qur'anic revelations, and to accommodate the new evolutionary words, terms, phrases and God to man inspired intermediary and awe-majesty: the Arabic tongue and language was radically changed, revolutionized and expressly transformed. The Arabic language derived from the Semitic Aramaic, spoken by Jesus Christ. More conspicuously of interest, was that linguistic Aramaic embodied a greater expanse of mystical, spiritual, and aesthetic Bedouin desert expressions than Hebrew. And thus, the tradition of paragon, that "Arabic is spoken from the heart". Even the sound vowels and utterances follow the human natural feelings and states of mind and heart.

Actual words depict the shades of being, such as sadness and tears, by "bi'bki", or laughter, by "bi'thaha". The voice itself expresses the mood. It has also been noted by linguists, that languages, indeed do express, not only the cultural perspective of a peoples or nation, but the psychic disposition. Similarly, the French language is spoken through the nose, and English, through the teeth. The subtle inferences comprise a whole subject of further prognosis and inquiry, not within our reach herein.

Needless to say, the Arabic language, with its endless manifestations of God inspired lexicon- must transcend audibly heavenwards... Infiltrate and penetrate the ever-living ether of eternity and infinity of being. With "Peace... Al Salam"... Heaven's own language of pure peace... Revealed by the Holy Prophet. Hence the Prophet's commandment to mimic heaven's pure awe-aura of peace to man, below the celestial canopy. How, pray-tell, does such an Arabic language, infused, elated, and elevated with such God inspired resonances confront our gross crass de-spiritualized secular lingo? And revitalize our spiritual apathy? For our own soul's and posterity's sake!

Christendom's dwindling numbers worldwide, are verily tell tale unassailable evidence of the decline of religion encompassing western and international civilization. Conversely Islam's numbers now exceed 2 billion, in fast acceleration growth. Islam's pre-eminent appeal, as reiterated, is Islam's direct encounter and holy intimacy in unity with God, sans institutional hierarchal litany, control or conformity - other than God's singular authority. Islam overtly confirms Jesus Christ's revelation that "the kingdom of God is within you." Who other than God, can lead, direct and impact one's humanity, in the here and now? And, lest we forget Islam's God exalted condition for man ... As "forged in God's own Nature, "Al Fitrah": the heart of the Islamic revelation, (Sura 30:30) that man is "saved", by virtue of his own God innate nature, endows and bestows God's "anointed man" with the potentiality for perfection. Higher, more exalted God identifying reality for mankind, in lieu of fearsome wrath of hell fire and mortal sin! While Christianity revels in 'sin-ridden' denunciations of man, Islam raises man to God's own infused identity of being, and potentially "sinless" condition! One does not have to blindly wait on the hereafter; God is here and now relevant to life's every moment, waking and sleeping, conscious or non-conscious, the one unique God cherisher creator of his beloved man creature! Hell and heaven are here and now; states of being, not places, hell's redeeming fires, purifying redemption. "Hell, "God's love in reverse"... "Heaven". Here and now, in God's bestowed blessings of love, buss, peace and all God's beatitudes of God's own divine attributes vouchsafed man! Intimacies of heaven, here and now; in a lover's consummate devotion, a soul's true friend and confident, a true brother man or sister woman's unswerving faith and sustaining support, or one's solitude, in the wilderness, attended by pristine beauty's inebriating rapture and God's inviting inspiring soliloquy of heavenly omnipresence.

So it seems all too apparent, nay illuminating, that world Christianity's ever decreasing numbers, vis a vis Islam's ever accelerating international, unequivocally evidences that religion, albeit Islam, still provides and elicits a hunger and unfulfilled yearning to know God first-hand, in holy intimacy of sanctity, sans men's organized hierarchal governing institutions and doctrinal intermediary and interventions. Islam may yet prove the religio-spiritual moral salvation of western civilization - in our bleak, de-spiritualized, demoralized degradation. Islamic sciences may yet resolve and atone western secularist sciences annexation of religion and God, and restore God and the whole integral truth of immutable reality "Al Haqq"/God to universal nature, rectifying and curing our diminutive mechanistic fractures. While there is no magic potency to cure America's drug and alcohol culture addiction crises that disrupts and deranges otherwise sound minds, causing mental anxieties and depression, suicide and death, Islam can reinvigorate and revitalize our spiritual capacities to override, and surmount our bipolar tensions and depressions. Islam's "Unani" Islam's whole soul, mind and bodily therapy can cure our cancers and degenerative diseases; and, perchance, God willing, the dire need to resurrect our western American religious-spiritual efficacious moral veracity, will quell and dissipate our nation's street crimes senseless violence, restore peace, equity and equilibrium to our neighborhoods and communities! Yea, how our once conscionable Judeo-Christian heritage must needs return to its founding democratic roots, under God, America's "Founding Father Creator", Protector, and everlasting Sustainer!

We, of western American humanity, suffer from spiritual depravity. We need our Muslim brethren to lead us out of the abyss of disparity and despair to higher spiritual heights of transcendent enlightenment, in our intuiting endless quest for self-discovery and to glean the innermost truth and absolute of our immutable reality "Al Haqq"/God. That man instinctively rejects his or her mortality, only confirms his God intuited endowed nature, "al fit rah:/God, through whom he, man shares eternity! Man's inherited innate God identifying nature: to reiterate, the heart of the Islamic revelation for all humankind! (Sura 30:30) Islam's elucidating "salvation", by virtue of man's own God-bestowed oneness of nature and being, reveals the unequivocal truth, that man is, in essence, sinless. "Sin" only exists, as ego error, vis a vis truth/God "Al Haqq". "Sin" is illusion versus reality. Al Qur'an defines sin as "temporal. Doomed to perish... Ultimately, unreality... For God is the only reality." ("Al Haqq")

We, of American humanity, are in a soular crisis of identity! Islam poses our last salving hope and 'saving grace' to renounce our Godless secularism, abandon and annihilate our false God

48

of "egolatry", worship of ego. Time looms imperative and crucial while daily, hourly, our demoralized American civilization crumbles and collapses, with endemic crimes, of unabated violence, destruction and death to our innocence! While vile pervert's rape, and assault helpless victims. While unwanted human babies are discarded in dumpsters, or aborted before precious life's gift of being, while greed and corruption thrive in every venue of social, political economic life. While our Godless medics' expertise fails to cure our cancers pandemic. Yeah, my American countrymen and women: we stand in dire need of Islam's God-integral, remedial enlighten-ment - Islam's medical science's the whole soul, mind embodied therapy to cure our cancers and degenerating diseases, to cure our mental imbalanced dysfunctional depressions; to empower the spiritual vital forces, the "Ruah" -"the vital spirit" of Islam and the ancients, an invisible vital inextricable integral mobilizer, revitalized renewal of the human immune system. Our western mechanistic mania must expand its materialist stifling constraints, and adopt Islamic science's "Tawhid" - the whole inviolable unity of spiritual and material universal material realities, and the indivisible oneness of God, man and cosmos. And know ye, that our incredible magnificence of universal life being. No random chance or science conjecture as "accidental coincidence", the grand slam insulting blow to human rationality and commonsense intelligence, moreover debasing desecration of man's "anointed" faculty of "reason, the first thing that God created for his man creature"! Yea, know, that all universal nature is governed by flawless intelligential laws, unalterable, non-conforming (to man) and absolute; and we, of homo sapiens, beholden to this infinite omniscience, "Al Il'm", are linked by reason of our own God-inherent, endowed nature, "al fitrah". Our entire highest and optimal well-being, destiny and survival, as a species, depends on recognizing our latent, ineffable oneness with our "Who's Who" universal Originator, Genius Conceiver, infinitely beneficent creator in our 21st century "soular crises of identity"! And thereby, to evolve our God-potential for perfection, in loving unity and peace within all universal humanity.

God willing "In Sha Allah"!

CHAPTER 2
"THE DESCENT OF REASON IN WESTERN MAN"

"The first thing that God created was reason", reveals Islam. "Reason" - "Al Farqa'n", implanted and improvised by the Divine Mind for man's cognitive facilitating faculties. "Man", "El Ensa'an-in Arabic, translates "the knowledgeable creature "Reason", the holy human property and power to synthesize and analyze, decipher the "golden mean", and the insightful eye between truth and falsity, cause versus effect, conjecture, and mere supposition to concrete factuality. "Reason" that discerns the ephemeral mutable, from "Immutable Reality." ("Al Haqq"/God) Man can veritably reason his "Cause Creator", vis a vis the "created" man being. Or "Did ye create yourselves, O man?" blind faith ranks as abomination in Islam and grievous insult to the Universal Mind, "Al Aqal" - Who bestows man's own intellect and uniqueness of genial kind! Reiteratively, to decipher and differentiate between the genial rational minds of luminosity, versus the irrational darkling morass of nonsensicality that inevitably borders and beclouds sanity. Aye but let us celebrate the infinite bountiful beneficent creator of man's definitive divine reason.

"Praise thy bountiful magnanimous creator,
Who created man from a clot?
And taught man, by the pen,

51

What he knoweth not"!
(Sura 1, "Al Alaq" - "The Clot")

The al Qur'anic Sura, "The Clot" further reveals:

"Nay, but man is surely
inordinate, because he deems
himself as self sufficient."

Man's inordinate egoist self-sufficiency, was
seemingly far removed in the Holy Prophet's
epoch, than our own modern neo-religion of
"egolatry"- worship of ego, that my 21st century
peers espouse-albeit unawares; that beclouds,
shrouds, confuses and contorts, and yea-desecrates
the holy faculty of reason! Ego that ellipses the
perennial light of Truth, and Immutable Reality,
"Al Haqq", divinely innate and accessible to all
universal mankind!

The new "ego-tarian expertise... "egoist
expertise", "egoism", or call it what you will, bases
all knowledge of facts on the ego's lamentable
limits and superficial reasonless consensus:
electing fallacious academia and science "expertise
consensus" over the veracious intuitive light of
truth. We now have egotistic professional
autocratic academia elite who suffers from
pedantic dementia, supplanting our emerging and
illustrious metaphysical genius giants! And our
scientists' egoist expertise, who confound facts
with facetious, fictitious, meaningless 'ass-end'
morass data base and void, eclipsing the
underlying truths of universal nature and being - a

soulless, Godless materialistic malaise of nonsense and no consequences! Ad-lib, our science medics' ignorance' and abysmal failure to combat and cure America's tragic disease endemic! These medical quacks, who have failed to discern the underlying reasons of cause vis a vis the fatal diseased effects. Our medical "expertise", who abort, disdain, and disinherit the once whole scientia of man's spiritual corporeal persona being of man's inherent God "Omnisciently Perfect Physician" within man's soul, mind, and bodily embodiment man's invisible anatomical "vital soul" - "Ruah" of ancient medicine, verified by Islam's whole medical science of "Unani" - salving remedial cures for our western soulless, Godless sciences of obtuse, material, mechanistic intransigence and dross, endless datum going nowhere!

So, is reason nearly extinct, among western modernity's stifled mind-set? Has reason so devolved and descended among our scientific peers and practitioner leaders? Has reason so lost its intuiting underlying truths, in the heart and mind of contemporary western mankind? And ego and "egolatry" overrode our higher scan transcending the secular mundane, to our innate nature's compelling God-inborn, and endowed being, "Al Fitrah"? Then above all else, we must return, rebirth, renew, re-spiritualize and revitalize America's founding faith, under God, endowed with God's divine attributes of life, peace, liberty, love and brotherhood.

Oh, my 21st century American brethren:

53

lose thy ego, so constrained finitude for the infinite mind ("Al 'Aqal") of true enlightenment of Immutable Truth and Reality "Al Haqq", to the God originator of all knowledge and science. "Al 'Ilm" for such as we "non-self" created humans, are nothing in ourselves, bequeathed infinite stages of evolvement and enlightenment by our Beloving caring Beneficent Creator!

"Earth is the first growth, and there are eons of stages of evolving ascent truth-wards. God-wards... Beyond earth's initial being"! Reveals "Al Hadith" to all humankind.

Rene Descartes' famous medieval dictum - "Cogito ergo sum" - "I think, therefore I am", pronounced a reasonless utterance, linking the human rational mind, to the actuality of being. Whereas, had reason prompted and premised the medieval thinker, he would have reasoned and employed cause and effect. "I am, therefore God exists" or "I am, therefore, my being verifies God!" This more rational declaration of being would have resounded, with the light of intellect's spiritual eye of insightful truth, acknowledging the cause creator and life originator, "Al Hyatt"/God. God can be reasoned "creator" vis a vis man, the "created". God can be reasoned without religious presumptions, proscriptions or theological dogma. The "I am" created mortal, must necessarily attest a creator originator, who produces the likes of thee and me! One uniqueness of genius conceiver, who shapes our human marvel of soul, and intellect embodiment endowment "ye, 0 man, are my greatest miracle... Why look ye for lesser marvels!" reveals God to all mankind, from Islam's holiest of holies, "Al Qur'an".

The Latin Descartes premises "ego" on the forefront of western physical limiting mind-set, not metaphysical, spiritual, limitless, genial light of intellect, as propounded by the ancients and Islamic scientific enlightenment of truth, "Al Ilm", "metaphysics", as explored and attested by Schopenhauer, Goethe, Toynbee, Emerson, and other western geniuses, gleaning the light of universal truths, now besmirched, desecrated, profaned and reduced to mere "probabilities", "random" and "accidental" consensus of our cosmos magnificence, and flawless order, and denunciations of metaphysical and physical unions of heretofore scientific axiomatic truths.

Descartes opened the door, forever eluding scientific Truth and Immutable Reality, "Al Haqq"/God, to the "relative" empirical secular literary and scientific protocols, such as Hume, and his egoist "humanists", and Darwin, who premised relative factual datum as the basis for scientific bogus truths. The universality of truth was ostensibly compromised for fallacious meaningless voids relative to the "relative", not relative to "the absolute".

The evolution and development of earth's land and marine animals was studied and recorded centuries before Darwin, by Islamic biologists, zoologists and geologists, whose discoveries and conclusive truths differ at converse odds with Darwin. All animals were endowed with intellectual instincts, for their protection, preservation and survival, not by some vague "natural selection" mechanism - which eludes nature", by definition, as to what "nature" implies relative to nature's overall role in harnessing the selective behaviour and development, and to avoid any inference to universal nature as originator, provisionary, and higher empowering, overruling life-sustaining, creating God factor! Darwin's other supposition, that "the survival of the fittest" ensures evolutionary posterity, once again is at odds with Islamic zoology, whose science expounds that only by cooperation, loving unity, and fierce protection for its species-kind, does universal nature animal beings thrive and prosper successfully. Try disturbing a beehive or ant colony, and witness their fierce, instinctive, and protective love arsenal for the preservation of their families! The Darwin theory for evolution is mere observational material mechanics, devoid of reason, altogether, in attesting effectuality to causality and denial of universal nature's omniscient overseer, genius conceiver creator! Reasonless, in discerning the spiritual abstract and concrete realities, in perfect complement and concordance in defining the whole science of life being. The art and science of reason, never exercised by Darwin, and later posthumous western scientists in their crass, gross, corporeal blindness, and fallacious suppositions, conversely Islamic science never witnessed these disparate fractures of the holistic visage of universal nature.

In the "Al Batin" metaphysics of esoteric, spiritual, invisible, underlying realities, in perfect "Tawhid" unity with the "Al Thahir" exoteric physical visible realities.

Bones, fossils and cranioskeletal geologic remains, can in no way depict the evolvement and evolution of plant, animals, or the homosapiens species without the spiritually charged living essences that preempt and shape the corporeal forms. To reiterate again - only the whole science of universal living thing and being can sanely rationalize life's marvel of reality. The apparent dispute between Darwinian science and the Christian gospel is the western Greco-Latin bible's literal misconstrument, vis a vis eastern Christian Aramaic-Syria original sources, which manifest the metaphorical symbolic language so indigenous to eastern spiritual temperament. A "day" can symbolize a time span of ten thousand millennia, millions or billions of transpiring years, to produce the overwhelming awe-marvel and wonderment of life's universality of being. No less an incredulous marvel and miracle! So the bible exacts full concordance with science's archaeo-logical discoveries, however the skeptics and "non-believers" revel in their egregious irrationality.

The same eastern allegorical language is used in the Christian bible to describe the tree of life, whose apples were consumed by Adam and Eve. Islam reveals the tree, emblematic of the fruit of knowledge and conscious realization of life-being. There is no fall from the celestial heaven. No "sin" or "original sin" of any absurd kind marks mankind. God willed that man should embark on earth's journey of conscious life being from his or her unconscious celestial state of living, heretofore, in God's perennial spiritual essence of being. "Original good/God" preceded man's earthly consciousness. God willed man to "be"! He, the originator creator has only to say 'be'! And it, he, or she is"! Revealeth Al Qur'an. From God's Pure Spirit to spirit and flesh, inaugurated man's creation! Adam is the first God-conscious man-being, not the Paleolithic prototype freak that perished millions of eons before.

Al Qur'an is explicit in revealing man's origin. Composed of water, earth's pure air, elements and components, all enlivened and infused with God's ever living eternal spirit, imaging light of intellect, and infused and "anointed" with God's own innate identifying nature, "Al Fitrah"/God. "From a tiny insignificant sperm drop of life essence... Then, a blood clot... He shaped thee. It is I, your cherishing creator, who shaped you in your mother's womb, gave you seeing and hearing... Sanctified holy unions of love between thy pair mate companion, and family!"

"Was there a time when man was nothing to be remembered"? God reminds man, in "Al Hadith". "Nothingness... To... Some one uniqueness of man being!"

"I was a treasure unknown. I created man, in order to be known," by knowing God, man comes to know himself, and by knowing himself, man comes to know God - his "Who's Who" ever beneficent, creator cherisher. For the first, unprecedented time in religious history, man's evolvement is 'a priori', inextricably linked to the knowledge and realization of his life's originator creator; for the first iconic time in world history, the perfecting evolution of the man-species is altogether contingent on his scientific enlightenment of truth, and immutable realty of being, "Al Haqq"/God. "For God is the only reality!" reveals Islam. God's exigency henceforth mandates mankind's everyday societal, cultural living reality and to wit, futurity and immortal eternal destiny!

We, of humankind, endowed with rational reason, are importuned by our reason's provider: "se ye, and witness the world's cities, communities and civilizations of peoples that once thrived, flourished, and have since perished. Only those, whose certitude of good, moral veracity, kind deeds and loving service to their brethren endure, in their remembrance of me God, their creator, perpetuate by their posterity."

We moderns have only to behold the pagan Greek and Roman ruins, albeit all are accountable, good benefactors and evil perpetrators alike. Apparently the moral veracious fiber preserves civilizations from total extinction.

How do we weigh and assess the currency of Islam's international nations, and communities of some 2 billion, whose posterity have perpetuated over one and a half millennia, where God permeates everyday social cultural living reality here and now! Islam, past and present, divulges an anomaly and unrecognizable phenomena, vis a vis our western secular God disparate civilization!

Islam never suffered the west's fatal science schism betwixt religion and God: in fact and reality, Islam ignited the scientific search for enlightenment and universal truths, whose original scientific discoveries and colossus of advancements in physics, astrocosmology, and mathematical sciences, founded the basis for western scientific technology. Islam had, never constituted religion, per se, as other world religions, but the science and nature of universal living reality, in flux, here and now, and incorporating the principles and immutable Truths of Perennial and Perpetual Reality - "Al Haqq"/God. From the start of the Islamic revelation, the holy prophet equated the quest for knowledge and enlightenment as the science of immutable Truth: "A Divine commandment incumbent on all Muslims... From the cradle to the grave"... As the pre-eminent act of salvation of soul, mind and heart, in pursuit of enlightenment of man's own God imbued Nature and Being - "Al Fltrah"/God. And conscious attainment of peace and ineffable union in the infinite all loving, cherishing creator. Truth leads ultimately to love's oneness, jubilance, bliss and peace. "Al Islam": God's name for "peace", Islam is to sweet surrender your self-ego to "peace" and the infinity of all wisdom. "Al Il'm". "Earth, being the first growth and there are eons of evolvement truth-wards, of enlightenment, after transcending this world's domain."

We reiterate yet again Islam's whole science of universal life being, Islam's science of "Tawhid": the underlying esoteric invisible spiritual reality, "Al Batin"; in perfect compliment with exoteric visible materiality, "Al Thahir", which our western sciences are in so dire need of, with its fragmented, decimated material matrix. Western astrophysics premises a cosmos devoid of any spiritual energizing causal creating power originator, which forms and shapes material objects; reduces the universe to "particles", atoms, molecules, gaseous entity, as constituting mere materiality... And gross physical intransigence, obliquely all living spirituality. Western science of astrophysics confesses the "invisible matters" existence, and concedes dark matter and black holes, however inexplicable. How be it rational reason, does not resolve the perplexing mystery, and conclude, the reality of spiritual forces, governing all physical phenomena, and the cosmos originator, cause creator, the overall precipitating, energizing factor! 0, but this might shatter the unholy alliance science consensus of their Godless universe. Albeit the new charging life forces create ever new magnificence of constellations and galaxies; what insane, absurd irrationally forbids we attribute these, and all universal illustrious marvels to its namesake Genius Creator, who makes scientific process even possible. So ingrained in western science DNA, is this notorious absence of God from God's own marvelous miraculous cosmos?

Astrophysics' theorem of "random chance", "probabilities" and other denials of irrevocable evidences of the immutable creating God factor in the universe were denounced by Albert Einstein, in his famous declaration that "God does not play dice with the universe". Einstein's belated wish was that he failed to prove God, as indisputable creator factor in the universe. And in confirming the cosmos' perfection, God as supreme architect, supernatural engineer, mathematician, and all-omniscient all-knowing perfecter, of all mankind beholds in endless, breathless wonderment!

Let us, as lay scientists, examine science's indisputable facts comprising for example, precise distances between our solar star and the earth. How can one possibly rationalize or reason that the 93 million mile distance was perfectly timed, for man's appearance on planet earth? Otherwise the sun would have burnt the earth to scorched oblivion, prior to man's creation! Firstly, we know the earth had to cool, precisely to admit man to the exact temperature and ether that would sustain life, as we humans know it. And earth's moon and sea tides' orbital gravity, working in rhythmic harmony within our orbital sphere; and earth's all-powerful gravity that protects our planet from outer space debris, comets and asteroids. All, reiteratively, perfectly timed! Reason alone validates the glaring omniscient reality of the God absolute and uniqueness of creation! That Earth's gravity (called "ish'ti"- translated as "love"), or

God's protective 'glue' that holds the perfect balance between earth and our galaxy, "see ye any pillars supporting earth's buoyant suspense in the cosmos?" challenges the God architect of the marvelous cosmos! And what inexplicable marvel of man's incredible psychic mind and bodily miracle of being! All ye must do is fathom the reasonability of your own sanity, and commonsense intelligence to assess and pre-adventure, celebrate the arch-architect's artistry and engineering feat of God's inimitable wonders, versus man's frail creative ingenuities! Albeit, we are vouchsafed God's inspired creative joys and mimicking jockeys as aspiring would-be artists!

The "fly-by" Pluto flight undertaken by "new horizons" spacecraft, this July 2015 marks yet another milestone in NASA's discovery of another planet's related surfaces similar to earth's, and once again, our planets and galaxies time frame of the "big bang" birth of our solar system and universe. The photos taken by the spacecraft, revealed high mountain ranges and deep gulf crevices, and what appeared as ice lakes and possible water present on Pluto, and exciting speculation that the amino acid building blocks of life are present, and life exits in some bacterial or microorganism form. If so, astrophysics or physics might then, conclude that the same amino atoms formed life on earth; but the human organism complex still cannot be explained in terms of atoms, molecules and proteins. In western typical mechanistic materiality, minus the vital life forces of spirituality of man's substance of living being and reality, which goes amiss and inexplicable by our secular science's material constraints and woeful marginal limits' man's psychic, mind cognizant, and emotive properties, go wholly unanswered, only reiteratively, in techno-reduction, and soulless Godless fallacy. To rationalize man's exigent liaison to any 'who's who" originator genius cause/creator, would be blasphemous treason to science collective consensus of the status quo atheist expertise… who abysmally fail to enlighten the holistic living nature, and being identity of our phenomenal wondrous beauteous mankind! We, of man, are not atoms, molecules, nebulous blobs, but living soul heads competing from our life-charger, promulgator, spiritual fire-storm God epicenter "Al Ruah", and emoting from our heart's heart of sensual vibrancy of love being, "Al Rahman."

Oswald Spengler, in his severe critique of western science's hapless material conundrum, states in his "The Decline of the West" - "that nonsense of all nonsense within science is a misdirected attempt to deal mechanically with the living content of scientific knowledge". Spengler speaks further "of the materialist conception of history, which springs from the same root as Darwinism, and like it, kills all that is organic and fateful. Thus the morphological element of the causal is the principle and the morphological element of destiny is an idea, an idea that is incapable of being "cognized", described or defined, and can only be felt and inwardly lived".

The same baseless mechanical medical science pervades our medical quacks' failure to combat and cure our fatal cancers and chronic disease endemic. We are in dire need of Islamic medical science's the whole soul, mind and bodily therapy, exercised successfully over centuries. The human psychic-spiritual persona configures uppermost: the ace determinant governing the mental and bodily state, and medical deterrent for preventive illnesses and diseases afflictions. The "Al Ruah"/God "Soul's soul" is man's key, foremost uppermost empowering spiritual energizing, revitalizing sources and resources, summoned in prayer, transcending life's mundane prayer, in oneness of soul, intellect and heart "in converse with God" and espousal of loving repartee, bliss and peace. Thus, our western mental defections, disparities, and depressions are unknown to Muslims' safeguard of peaceful union in God, whose name is "Peace"... "Al Islam" ye have only to forego "ego", and solicit... Sweet surrender of all self and heart to the one beneficent, beloving creator! Man's soul, mind and bodily God inherent "Perfect Physician", whose divine embodiment resuscitates, renews, regenerates man and woman's "invisible" anatomy, and visible limbs, tissues, muscles revitalizes and restores the human immune system!

Drugs, prescriptive drugs, of western therapy have abysmally failed to cure mental and physical dysfunctions, disorders, and chronic recurring diseases. Clinical depressions can be prevented early on, with prayer and positive spiritual access God-wards "oh ye, of little faith". Faith, as such, is not religious indoctrination, but faith innate, within thyself, life's incredible miracle and gift blessedly vouchsafed you, and some one-life originator, benefactor - who made this whole marvel of being possible, my earthly co-habitant! How can this positive reality not render faith and unending gratitude to life and love's unsolicited giver?

Incidentally, after God's 'a priori' positive healing cures, this author recognizes the true genius of the science of psychology and psychiatric therapy. Carl Jung cognized the whole organic human spiritual, mental and physical enterprise over his scientific peers and egotistic, atheist conformers. Jung forewarned that man's salving "subconscious" God authoritarian and unalterable law of man's nature, if alienated or violated, by man, by spurious self-deception, and superficial psychic methodologies (Freud and his abject sex ideology) would exacerbate neurological-psychic mental illnesses and discords. Jung's genius of the inviolable "God self-consciousness" within, has prophesied our now soulless, Godless western egotistic atheist sciences that have only exacerbated modernity's neurological depressions and mental illnesses crises. Carl Jung would be flabbergasted, by contemporary psychiatric practices, whose reliance on drug therapies have grievously failed to remedy or cure, in any way, drug after newer drugs conspicuously Prozac, have resulted in tragic suicidal deaths. Eluding and eliminating altogether spiritual reliance and God rehabilitating, revitalizing resources within; that bequeath peace of mind and heart to troubled disenfranchised drugged patient victims, of yet more expansive experimental drug therapies that inevitably fail.

It was primarily the spiritual eye and abstract genius of Islamic mathematics that transposed Pythagorean pagan Greco-Roman numerals into the Arabic qualitative and quantitative symbols forming the whole integral summation of mathematical science, as we know it today. Pre-eminent, ever the God factor: symbolized by zero, or infinity, indivisible, intangential, altogether, greater than its parts. God/Infinity... Neither triune nor finite, as in Christian orthodoxy. One Asian commented that were it not for Islam's concept of God as infinite, we would not have mathematics, as we know it.

Thus from the zero axiom of God's infinite premise, derived the decimal system, and Omar Khayyam's equations of quantum physics symbols - making possible time, space and light years distances. Islamic mathematics dealt with realities, here and now, in flux; discarding and annihilating prior Greek theorems and philosophy, which failed to address living universal phenomenal being and realities. The mathematical sciences of Islam revolutionized forever mathematics, and evolved the technological leap forwards in astrophysics, physics mathematics, albeit unaware or credited by modern scientists. The symbolic spiritual essence of abstract Arabic numerals still eludes our modernists, who will never comprehend the religio-spiritual God symbol, underlying their techno-mathematical applications, apprehended as western science!

Al Ghazali, famed mystic, physicist, and meta-physicist, resonates the al Qur'anic ultimate truth of revelation: that "God is the only Reality" - "Al Haqq"; "reality as comprehensive of both mutable finite, visible and immutable infinite invisible spiritual realities. Al Ghazali verifies that nothing is physical, but has a spiritual originating spearhead in Truth, "Al Haqq"/God. All materiality exists, contingent on spiritual essences energizing and constituting being entity. Even inanimate objects such as seemingly lifeless table, chair, desk, or other man-made construction or invention, still contain the invisible minus and plus, electron-proton molecular substances that charge visible forms of concrete reality. The genesis of biological science, of minus and plus electro-opposite energies, discovered by Islam, was dismissed as "sensual sexuality' by the medieval church, hilarious as it may seem today. Fundamental constituents for electricity, Islam's unprecedented discovery, of the electron (minus) and proton (plus) chemical balance, was harnessed and engineered into producing the world's first generators and electricity in 9 A.D., lighting the cities of Cordova, Spain and other Muslim governed public provinces.

Since western Christendom had posed a hapless hiatus and infamous Cordova in secular societal life, and corresponding cultural scholastic denture, with the decline of spiritual religious vitality; the truly inspired poets, artists and literary luminaries in their endless quest for perennial truth and God, used metaphoric language and allegorical relevance to God and universal truths enduring ages on, and veritably reconfirmed that man's innate, inborn intuition for eternal truth and God, yet thrives and abides in the bosom of mankind!

Percy Bysshe Shelley, 'the mystic genius visionary, out-soars them all, transcending soulful meteoric heights, lamenting the tragic death of Keats, and mortality's doleful abyss. In "Adonais", Shelley pens:

"Life, like a dome of many colored glass,
Stains the white radiance of
eternity, until death tramples it to
Fragments."

He continues in later lines:

"Peace, peace, he is not dead, he doth not sleep,
He bath awakened from the dream of life.
'Tis we, when lost on stormy visions, keep with phantoms an unprofitable strife.
We decay like corpses.
Fear and grief convulse us and consume us day by day
And cold hopes swarm like worms within our living clay.
He has out soared the
Shadow of our night.

He wakes... 'Tis death is dead,
Not he."

Metaphorically, Shelley lyrically presents the perennial reality of Godhood, and the spiritual immortality we share with our eternal being.

To reverberate, these spiritually illumined minds, of whom I shall gloss over, later among western nineteenth century genial greats, wholly committed to humanity's higher evolvement and transformation, rebelled against the corrupt civil, cultural, political and economic status quo, and rejected western science's Darwinian soulless, Godless materialist matrix and mechanist stalemate, and fast eroding moral veracity. My space is too brief to name or address the other giants of intellectual genius and perspicacious insightful eye such as Goethe, Toynbee, Voltaire, Victor Hugo, John Locke, Carl Jung, Thomas Carlyle, and our American prophet, Ralph Waldo Emerson. (Thoreau's mentor) Emerson, formerly a distinguished clergyman, rejected the Christian tradition of doctrine "dogma", for the living transcendental God "over-soul" within man; he recognized the inviolable, unalterable laws of man's nature; what we call good, is in accord with man's good/God innate nature, what we call evil, opposes and violates his nature and himself. If Emerson resonates Islam's heart of revelation, of man's identifying God innate nature, "Al Fitrah"; it only attests that universal truth is accessible to all mankind. Follow, our other "truth-sayers" such as John Muir, Teddy, Eleanor and Franklin Roosevelt, and Rachel Carson - to name a few of our genial minds and faith abiding hearts, prophetic sages and veracious moralists, undaunted, non-conformists, who were not intimidated by the skeptics, pessimists and egotistic atheists. These few illustrious greats were never timid or half-hearted to dispel gloom and the nay-sayers, but to enlist and celebrate God's highest providence, in the affairs of' man!

"Reason, the first thing that God created for his knowledgeable man creature, "Al En'ssan", remains the holy grail of the human mind; that holy faculty that apprehends and discerns truth from falsehood... That leads God-ward, from the finite shadows of earth's mutability to the infinite imaging light of eye and Truth of Immutable Reality. ("Al Haqq"/God)

"Earth is the first growth, and there are eons of evolving ascent truth-wards and Godliness, in pure bliss and peerless peace"!

CHAPTER 3
"THE EVIL JIHADIST FARCE"

"Jihadist", as in "terrorist", whom we identify as barbarous demonical enemies of humanity: Muslims, non-Muslims, Christians, Jews, all innocence alike: is, in historical fact and reality, an anomaly in the fifteen plus centuries of Islamic civilization. The very word "jihad": belies the Qur'anic holy revelatory meaning, of man's struggle God-wards for spiritual purification and righteous acts of loving service to God and all universal human brethren. See ye, 0 reader-why the terrorists "jihad" is an outright "farce"! These evil devils dare defile, desecrate the holy name of Islam... "Al Islam"; God's name for peace! Among God's other names for "justice", "Al Ad'el", liberty, "Al Thahir", "the infinite beneficent, all beloving, ever compassionate, merciful", "Al Rahman, Al Rahim"! In the holiest al Qur'an, God explicitly "condemns the aggressor"... "War is forbidden, except in self-defence... Pick up thy sword and defend thyself". The evil slaughter of defenseless innocent lives, constitutes a crime against "El Islam"/peace, God and all humanity!

"To harm one innocent human being, is to wage crimes against all humankind!" expressly warned and admonished are "the evil criminal perpetrators" who become fuel for hell fire! There is no way, these hell's devils can possibly misconstrue or rephrase this holiest Qur'an's primal revelation; there is no way, in the holiest name of the "holiest infinitely all-beloving, compassionate, merciful."... "Al Rahman, Al Rahim", no way, these accursed evil-doers can delete God's unalterable, and immutable truth! However they falsify their Qur'anic quotes and desecrate by outright lies!

Within the present-day global community of humankind, "Al Dar" of some 2 billions populace, peradventure 1/16th of one percent constitute existent "terrorists"; we reiterate Islamic religion infused pervasive as a commonplace moral veracious living societal, cultural unification of peaceful non-ethical, non-racial egalitarian equilibrium of human brethren, Muslims and non-Muslims alike in an unprecedented humanitarian success, never achieved in world history, ever since. This Islamic dynamic of egalitarian brotherhood is rarely known or realized by the media critiques, who lump "Islamic militants" and demonic terrorists in the same ignoble notorious glaring omission of the historical facts, past and present. "Umma... "Brotherhood" is a living everyday reality in social, communal interaction, in Muslims' DNA, enunciated in ordinary daily converse, greetings as "A'hi", brother or "Ich'ti", "sister". The holy prophet of Islam forever resonates that "humanity is one family". Of course, if the ignorant outsider has not visited nor lived amongst the Muslim community, we can make allowances for his abysmal ignorance, unless, he or she is blissfully blind-sided by his or her egoist preference of ignorance over truth and contemporaneous reality.

There was no "jihadist" terrorist, "Islamic militant" in our recent rhetoric of terrorists- "al Qaeda", "Taliban", and the other offshoots, of demonic hellions, et al: until, we, the United States of America, invaded Iraq. Until we, the people, and our war hawk policy leaders, succumbed to false propaganda, and willfully lied by fallacious intelligence. Neither was the 9-11 tragedy justified for waging war on Afghanistan. Both wars wrought death, destruction, and displaced millions of people, creating massive refugees from their homes, cities, and communities. There was no "Sunni", "Shieh" distinction of sectarian division, until our U.S.A. became involved in Iraq's internal affairs, and set up a "Shieh" prime minister to represent the Shieh minority, excluding the Sunni majority, and causing catastrophic political, social disparity and disarray; and casting Iraqi civil society vulnerable to terrorists, Al Qaeda and the Taliban, who seized their wicked opportunity to inflict chaos and carnage on helpless and defenseless civilians! Had America and our western partners stayed out of the middle east altogether, the people themselves would have resolved their political, legitimate rights to social, libertarian and economic justice, to rid them of the tyrannical injustices of Saddam Hussein. For those of us, infinitely grateful for democratic freedom and equality for unanimity of life's sacro-liberties of intellect, person and property, we only naturally attribute the same rights of self-determination to all humanity else-where to aid peoples, in their legitimate causes and God-given aspirations for our own precious liberties, is certainly one of human brethren compassion. Such legitimate humanitarian aid, can take the passage of dialogue, diplomacy and delegates of peace, justice and liberty; as well as the last resort to providing arms if necessary to combat terrorizing civil warring injustices and prevent human carnage, and to vindicate capital, moral support and humanitarian stamina, by any and all means possible. This we could have offered, on behalf of the Iraqis suppressed and subjugated by Hussein's evil regime. Besides the new tragic refugees homeless disaster, we Americans lost our own precious American soldiers lives, wholly unjustifiable, moreover, their loss of life and limb, sacrifices, sheer senseless tragedy, that could and should have been averted, at all costs whatsoever!

Syria is a whole different matter. The conscience of the international community of nations remained paralyzed and ineffectual, while Assad perpetrated genocide on his own people, destroying whole cities, of millions of defenseless Syrian civilians with relentless barrel bombs and targeted rockets; helpless children, women and entire families suffering merciless carnage! Some 50 million refugees have traveled hundreds of miles, sometimes on foot, to secure refuge in U.N. makeshift camps, or wherever Lebanese, Jordanian, or Turkish patriots for justice, provided shelter for these tragic victims. Yea, my fellow humanitarian Americans, we could have given the free Syrian army and other anti-Assad rebels, the weaponry and vital military, medicinal and survival sustenance, so vitally needed to combat and overcome Syria and the world's worst evil of dictator devils since Hitler! We would not risk the blood of American veterans, but how can human conscience not avert and repel the rapacious slaughter of millions of defenseless innocents, and repose calmly day or night?

Our absence in the Syrian tragedy allowed Isis to enter the terrorist marauding war arena, infiltrate and besiege Iraqi, Syrian and Turkish borders. The evil Isis savage brutality knows no bounds, in their "devil"... "Shatan" personified of barbaric heinous hellish merciless carnage of human innocent lives - no matter, Muslims, non-Muslims, Christians, Jews, of all ethnic and religious kinds alike! "Khalifate": a preposterous lie of all fallacious lies!! Islamic history produced Khalifate of unrivaled virtue, and governance of social justice, equality and liberty for the common man, be he Muslim, Jew, Christian, or non-believer atheist! These devil Isis, are all psychopaths deranged mad-dogs, incensed by power wielding crazy inhumane devilry! The real Muslims, Iraqis, Kurds, and now Turks will inevitably win over devilry! God or good, inherent within man's own nature, "Al Fitrah": will ultimately triumph over all evil. "For evil-ephemeral in time, perishes as unreal"... Revealeth God of all mankind... In al Qur'an. For "God/Good is the only reality!"

Suicide is strictly and unconditionally

forbidden in Islam, in any shape, form or act of justification. You dishonor your maker, life benefactor, and deny your God sovereign creator, and future ultimatum of immutable destiny. "Your return, oh man being, as all creation, is to me!" Now, enter the suicidal banner, under what all terrorists operate; bombs exploding, in vehicles, or on one's body by suicide, destroying human lives and wreaking havoc, in the name of the blessed God of all created life being! What God, pray tell, would sponsor such evil heinous brutality of inhumanity? What God, of any religion would rain terror, death and destruction on innocent defenseless victims? Preeminently, Islam, which forbids the killing of one human life being, as quoted aforesaid in the Holy Qur'an, which tragically, has escaped our terrorist killers. Fear not, hell fire awaits them, yawning to consume their evil "haram". Who desecrate the holiest name of "al Islam", "peace", or God's foremost exalted, holiest attribute, vouchsafed man, as "trustee" and holy "guardian of his brethren man!"

The "recruits" that Isis attracts come mainly from western countries, not the Arabic mid-east. All become ensnared in Isis propagandist lies, and "redemptive" illusions in the Isis cause of suicidal murder mayhem of human victims. Without exception, all recruits, either have a criminal background, are drug addicts, suffer from mental deranged illness, psychopaths like Isis psychopathic maniacs; desperados and degenerates, marginalised and disenfranchIsed, who come from the lowest bottomless bowels of human society; without family, friends or consoling support... Easily lured into the ISIS "religious fascism" these poor piteous "losers" have nothing more to lose; so why not take their own "hate spate" out on the human race that have abandoned them, their rightful place and coveted distinction! These pathetic duped fools are ready despoilers, in the "ISIS killing machine", unprecedented throughout human history!

For the immediate remedial and realistic solution possible, to this anti-Muslim, anti-human, Isis evil jihadist farce fascism: is to wage real "jihad", by the American and international Muslim communities, in a counter-propaganda war: against the rabid mad terrorist enemies of Islam and humanity! Islam's millions of ready recruiters from their vast international communities, "Al Dar", from the common Muslim man, and from Islam's eminent dedicated sheikhs, imams, education leaders, and distinguished heads of renown universities, such as Cairo's "Al Azhar University" and throughout the mid-eastern, African Asia and European Anglican nations. America's Muslims of 30 million plus-can readily, first-hand recruit the same Islamic distinguished academia, heading Islamic studies, in all American universities nationwide, while engaging common Muslim citizenry. A great majority of our American Muslim emigrants are proud, unconditionally loyal to our American democratic republic, many had been persecuted by their former Turkish ottoman oppressors and colonial occupiers; and in the Islamic religious context, Jesus Christ messiah is venerated; so emigrating to our Christian-founded United States democracy, was in perfect accord and harmony with Muslim precepts, Christianity and unanimity of moral veracity.

Mosques and Muslim communities will pose eager to assist such a counter-terrorist assist movement and eventual transformation to repel and expel the demonical Isis recruits from committing crimes here in our United States of America.

O, my fellow Americans, we too have our share of domestic terrorism. Crime stalks and terrorizes our streets daily with senseless violence, destruction and death. Our children are not safe in their schools; mass killings occur in our theaters, college and university campuses, and public malls and places, with no refuge or safety. Rape ravages every sector of living society-and murder accelerates daily, and hourly! The same psychopath deranged criminals roam and "own" our homes, streets, public edifices and institutions... Nothing and no one safe no one being, who is not vulnerable to criminal onslaught and life-threatening random abuses, attacks and mayhem murder! Can we not plainly see and acknowledge America's devastating moral degradation, from a once conscionable nation under God, with moral veracious, religio-spiritual vitality? The author has opted for a rebirth and spiritual resuscitation resurgence and renewal of our religio-spiritual moral veracity, under our God founded, impacted moral humanitarian democracy!

Let us all pray, "Insha Allah", "God willing", that eastern and western humanity alike, serve God, our infinitely beneficent creator, of all universal being, in true loving brotherhood equanimity and peace!

CHAPTER 4
"Unani", the whole soul, mind and bodily therapy of Islamic medical science

We Americans are the sickest of all nations in the western hemisphere, and amongst Mid-Eastern, Asian, international populations. American humanity is perishing of cancer and degenerating diseases in pandemic droves. At this timing, the year 2017, cancer fatality is consuming one out of every 3 lives, and accelerating daily. Western man is the world's most endangered species! The "cancer terrorist" within threatens and outweighs all of life's extraneous hazards, albeit America's street crimes senseless violence wreaks relentless danger to our everyday citizenry.

Our medics' "expertise" abysmally fail to heal, cure and eradicate this deadliest of diseases, that ravages and destroys so many precious lives- and now, even inflicting the pregnant mother's womb, endangering posterity. Our medics treat cancers and degenerative diseases, rather than decipher and combat cancers underlying causes. The sham of western medical science collusion in their "science and religion schism": precludes the primal preeminence of man's spiritual persona and vital soul veracity; that empowering regenerating, revitalizing impact on the mind, body, and human immune system! We suffer first and foremost from spiritual depravity. Conversely, Islamic medical science recognizes and prioritizes man's "vital spirit". "Ruah" as the pivotal over-powering impact on the entire human organism; and on its spiritual equilibrium, harmony and peace.

Our soulless, Godless "science schism" cult, and spiritual depravity, has cost the human patient's adverse de-spiritualized diminutive devastating loss of spiritual veracity and vitalizing powers to transcend the "religio-spiritual boycott" as some unreliable province of religion, not science religion and God, are taboo, in the mechanistic transience of secularist western science, et al.

"Unani" - the original whole soul, mind and bodily optimal health therapy of Islamic medical science can realistically offer the exigent alternative to western science's lamentable soulless, Godless cureless cancers and degenerative diseases, heart vascular failure, rheumatoid arthritis, diabetes, multiple sclerosis, liver, colon and all vital diseased organs, et cetera, the ever new recurring diseases, and their affiliate new viral infectious malignancies exacerbating daily.

Islamic science belies the western science/religious "hoax" - as we have reiterated; Islam, in actuality, ignited the search, discoveries and radical advancement of scientific truth, in all universal natural phenomenal being.

There is a "genealogy", of spirituality, unrecognized and untapped into, by western science. Spiritual genes, whose revitalizing forces overpower all other human immunizing factors albeit, diet consumption and physical exercise play a minor role. If the human mundane pray ten times daily, and ascend God-wards, "to converse" with one's beneficent creator; is it any wonder, that spiritual veracious essences regenerates and re-energizes one's human DNA condition of vital spiritual endowment of health and exquisite equilibrium and peace! Hence, spiritual resurgence on a daily basis has proven the true "elixir" of life renewable forces, against mental depression, emotive behavioral anxieties and dementia deterioration. These acute depressive factors are non-existent among Muslim kind. For God's consistent inner living affiliation, and loving oneness of union are ever-present! And lest we fail to mention-Muslim family consolidarity does not place their cherished elders and parents into rest homes to perish untimely tragic deaths! Moral veracity governs every facet of living, for Islamic civilization endures today, as the world's singular God-oriented population! What is normal for Muslims is "paranormal" to outsider observers what is natural to spiritual-tempered Muslims is "supernatural" to our western audience!

Heredity has proven genes are uppermost; musical genius exhibits forebear's musical preponderance, mathematics, and mechanist skills requires invention prowess; and as corresponds to spiritual propensities amongst man-spiritual inherited DNA does, indeed manifest the odd spirituality of some individuals over others, enjoying exuberant health, bliss and peace. To yet reverberate: that God and spiritual veracity are exigent to perfect health of soul, mind and body's human immune system!

Alcohol, amazingly either ignored, or unacknowledged by our medics is the key culprit causing cancers and degenerative diseases; "Alcohol's", derived from the Arabic, the greatest irony of all; was and is prohibited for its devastating destruction of billions of neurons, vital organs, heart valves, vascular blood flow, liver, kidneys and the entire deterioration of the human immune system. The vital liver's wonderful cleansing and detoxifying power-thwarted and disabled by alcohol's paralyzing assault. The human entire sympathetic nervous system and endocrine system are disinfected together with the stomach's alimentary canal contracts, and colonic intestinal excretions. So totally devastating are the deadly effects of alcohol consumption on the autonomic neurotransmitters, "voices" flawless intelligence governance of the wondrous human immune system! Islam's alcoholic discovery was used as an antiseptic purifying, sterilizing substance in laboratory research and surgery; drugs, poppy, cannabis or coca were utilized solely for their medicinal healing properties. Alcohol, the deadliest of all drugs, was and is still, strictly prohibited, for the aforementioned ensuing catastrophic destructions on the neurons, healthy cells and the vital protective immune cellular defences. While drug prescriptions originated in Arabic Islamic laboratories, prescription drugs were derived and extracted from living plant and herbal sources; and administered individually to each patient's unique needs.

A far cry from, and unlikely drugs experimental utilization by our western medical physicians, has ominous and oftentimes deadly side effects. Yet, how do we re-educate and enlighten our western and American brethren, to renounce and reject all alcoholic consumption, so deeply rooted is our European Anglo-Saxon habitual doleful "waste" scenario? Unless life and limb are prized above all other conformist importunateness, as we ourselves choose, as true masters of our fate and destiny! We, ourselves, "will" our own heaven or hell, create our own devils of disparity, or summon the angels of our own God intuited being and true identity!

The human soul is not some amorphous, ambiguous entity, the province of religion, as in our western science religious-schism: but life's overall generating factor and actuating substantial reality. Soular power that impacts the entire breadth and scope of the human corporeal embodiment. The body is sick, because the soul wanes sick. The body is beset with sickness, because the vital soul "Ruah" loses his or her essence of soular vivacity. The soul ("Ruah") is sick, out of tempo, in disconcerting disequilibrium and disconnect with the Soul's "over-soul... Al Ruah"/God. And the affiliate God infinite resources and all empowering life regenerating, revitalizing healing and curative agents or (angelic

85

forces that attend man's soul and bodily circumference) flawless intelligence that governs man's invisible spiritual infinite realities. Un-apprehended or detected by man's microscopic lenses, these hidden health immunizing forces, "obeying" in collective concert "the Perfect Physician within", in so obvious caring "TLC", every waking, sleeping infinitum moment; serving man's royal endowment! Unconscious, or consciously linked to love's holy "Intimacy of intimacies", love/God heals, cures, regenerates, and recreates optimal soul, mind and bodily exhilarating health, joyful exuberance, peace and equilibrium! "Unani", Islam's whole soul, mind and heart embodied medical therapy harnesses, and inspire direct link and liaison with God/love's ineffable unity and oneness. For to yet reiterate, love ultimately heals! Love's infinite beneficent, All Beloving, Compassionate Creator, "Al Rahman Al Rahim"!

Ye need only cooperate and "sweet surrender" our egotist self-constraints to our Genius Conceiver cherisher, originator of all knowledge, science and wisdom. "The infinite all knowing", "Al Il'm", and peradventure, glean some minute wisdom for our own human, God-created marvel of cognizant being! God resonates in God's revealed al Qur'an: "Ye, O man are my greatest miracle!" Shall we, created mortals, not give our life; love Creator, some credibility or acknowledgment?

The famous Ibn Sina (Avicenna)'s whose

"Canon of Medicine" was used by medieval Europe, as their medical bible 700 years introduced "the vital spirit", soul or "Ruah"; as the foremost determinant of man's mental and physical health status; and accordingly diagnosed all metabolic processes, as pre-essentially derived from the human soul, and vitalizing, energizing spiritual vivacity, empowering and impacting man's entire psychic bodily organic immune system; that the reverse, de-spirited, disparate, demoralized individuals, effectuated all illnesses and diseases, be it mental, emotive heart, or vital organ failure. The illustrious physician heralded that homo sapiens has his own inviolable nature's ecology and balance of spiritual, intellectual rational, corporeal organic phenomenal complex, inseparable and one with universal nature, and embodying man's own uniqueness of man nature and being, "Al Fitrah": his God endowed and innate nature. (Sura 30:30) Ibn Sina, together with posthumous physician doctors and researchers explored, tried and tested patients with spiritual veracity, as opposed to less spiritual tenacity, of lesser faith exuberance; and determined that the former spiritual volatile veracious, were far less prone and vulnerable to depression, and disease.

You must oust yourselves, my fellow Westerners, from your obstruct "prison box" liberate yourselves, from your own self-impaired constraints, and God/religious phobias. God is not "religion", per se, but integral to the whole of man's spiritual, intellectual, emotional and physical vital life being!

Who is the "I", in the "I am", your and my soliloquy of being, brethren man? Assuredly, "I" am not my muscles, tissues, organs, moving parts, limbs etc., but integrated one... With the "I am that I am"; the "Who's Who" Life Originator, Creator - "Al Hyatt"; whose spiritual living epicenter veracity, empowers and permeates every fibre, neuron and muscle of the human anatomical being; by unseen life-charging energizing forces that magnetically orbit man's unique ecology. Altogether mysteriously eluding our medic expertise, who see only the crass physical "incarnations" of mankind's astounding brain-body marvel of flawless functions operation! Islamic medical science gleans "the natural" as "supernatural", here and now in the present real, co-existent and contingent on "Immutable Reality", or "Al Haqq". You need no religious doctrine, belief or ideology, when man's psychic, mind bodily marvel singularly and unequivocally evidences concrete reality of an un-creatable supernal genius life conceiver, creator, who forges the likes of thee and me! "I am closer to you, than your jugular vein"! Reveals the Singular Originator Creator of man. "Why look ye, in vain for miracles other than yourselves? Ye are my greatest of miraculous marvels!"

"Unani" medical practice recognizes the alliance of the oneness of God and man: in the "I am that I am" unity, that releases revitalizing resources that superimpose vitality of psychic, mental and emotional health optimal resurgence and exuberance; you have only to forfeit ego, sweet surrender your self-egoist barrier, to the "Al Islam" of ineffable unity, oneness and peace. Be it a headache, spinal back pain, leg or limb sprain injury, mind, bodily fragile imbalance, etc., trust "the Perfect Physician" within, to combat and overcome all disablement and disorders. Surrender all human frailties to "Perfect Wisdom", "Al I'm" become whole and one in the "Infinite One", "Al Wahad". "Thy faith hath made thee whole", nothing is physical in itself, but all physical mutable matter emanates, existent and contingent on procreating spiritual Immutable Reality, "Al Haqq"/God. Universal nature's inviolable law of "Tawhid" indivisibly unifies both spiritual and material realities, as in-separately one in unison of visible exoteric and invisible esoteric ecological balances. Let us, at long last, seize some small, however meager glimmer of scientific enlightenment, of the spiritual invisible God Immutable Reality, beyond our microscopic lens and mechanist visibility that "glues" together protons, neutrons, atomic molecules and their hidden spiritual nuclear substance, initiating and materializing all visible universal thing and being. The "God factor" is the procreator of all living phenomena, and ultimate of all deceased life extinction, from our finite mortal plane. Let us endeavor to probe, reason and fathom, beyond our so limited finite myopic self-imposed blind sight, the limitless yea, infinite horizons, over and beyond our so constrained science consensus dead-end datum! Albeit, we will never quite glean the greater mystery, "magic" and marvel of life's incredulous miracle from nothingness, formless, spatial-less, timeless absolute irrevocable and Immutable Reality "Al Haqq"/God that forms, creates and substantiates our universal planetary plane as conscious reality to the likes of our human beholden! "I have only to say "Be" to it, he or she"! Reveals man's infinitely Beneficent Creator, the forever self-subsisting! Let us exuberate, celebrate and praise in our human evolving "scientia" true science of our incredible cosmos and us, "all praiseworthy celebrations" - "Subhan Nallah"!!

The human body alone provokes disbelief and marvel… Undeniable testimony of infinitude of Supernal Genius universality of Mind, "Al Aq'al"/God; witness the human corporeal embodiment; "concrete". Real evidence, irrevocable proof, with no religious assumptions or doctrinal thesis whatsoever! Physicians should be the most religious faithful of all, espying the mind neuro-bodily inconceivable marvels. But tragically, secularism has defiled Universal Nature's sacrosanct authority and uncontestable flawless wisdom!

Our frail reason fails us to rationalize or fathom the human body's preposterous wisdom of operation. How do the vital organs accomplish their singular functions? The extraordinary liver that cleanses purifies and detoxifies contaminants and pollutants? The marvelous heart, pulsating, man's vital breath of life-with no pause or intermission, during his or her entire lifetime duration? How do the twin kidneys precisely know when and what amount to discharge urine? And the intestines tracking and dissolving impurities with their acidophilus bacteria? And the bowels' anus, recycling the end toxic human waste? We humans, all too sadly, take the above miraculous vital functions for granted. Firstly, the incomprehensible intellectual laws that operate with precise perfection, and the so trans-parent loving care, exercised for our humanly highest and optimal well-being! Lest we forget, the human dental mouth, alone, stagnates all possible creative genius! Each single tooth has its own specific use; incisors, molars unique purposes, and acting in perfect synchronizing concert and harmony! Man's teeth complex proven perfect "pre-planning" ingenious conception! And what of our Love Overseer, to grant us the sumptuous pleasure and delight of taste, and palate!

We haven't even touched on the colossal of miracles, visual sight and audible sound! The astounding mysteries of the vision of seeing and auditory hearing still remain beyond man's cognizing scope of under-standing. While the ancient Greeks mechanically explained sight, as transmitted through the retina, Islamic optics discovered we see through our brain, which has since been adopted by modern science. Islamic science has more comprehensively combined the spiritual aspects inherent in the act of sight. That the brain images "the light of man's spiritual eye", seeing and mirroring the spiritual living essences of all things, objects and being, through that perennial and inextinguishable Light, "Al Nour"/God. Alike, man's auditory hearing, with its intricate canal system-emanates "spiritual waves" that intensify the epitome of seeing, with being! Still, my fellowman, we can never glean the marvel and miracle of our soul, brain and bodily marvelous being!!

"O man, it is I, your bountiful bestowing Creator, who has given you seeing and hearing!" (Al Quran)

We bipeds of regal stature are unknowingly, unconsciously, blessed recipients of our infinitely beneficent "Who's Who" originator Creator. Call it "grace", or rather God's unequivocal love towards the likes of you and me! Albeit the stark truth and concrete reality of man's flesh and blood, brain body being proven and per-adventure grasped, by the least reasoning and commonsense rational among us!

Love/God's beneficence supersedes all else, comprehensive to man. "Grace" translates to "Baraqah" in Qur'anic language: or God's love of infinite omnipresence, in oneness and ineffable transcendence of intellect, heart and soul in Union "Marifa" with God, instantaneous, here and now; not some afterlife quasi redemption. Love/God is accessible and infinitely irrevocable, here, within earth's time span, by virtue of man's God endowed and innate nature "Al Fitrah". Man is "saved", (in Christian lingo) not be church or clergy litany, doctrine or self-proclaimed salvation by whatever espoused proscription; but by virtue of his own God inherent nature, revealeth Islam, as we have reverberated, the heart of the al Qur'anic revelation. (Sura 30:30)

In experiencing the rapture of God's union of love-one Muslim mystic exclaimed his unbounded love for God, exceeded all his other possible virtues. But in re-examining his fervent claim: he perceived that love/God had preceded his own, even making his love possible! Another adoring Muslim, famous mystic, meta-physicist, and scientist, Ibn Arabi declared: "Love/God is the cause of all love. Were it not for love, God would not be worshiped!"

Love and immutable truth are always intertwined by the truly wise sages, as the famous Muslim Sufi, mathematician and scientific genius, Omar Khayyam said... "But while the eternal one created me, he word by word spelt out my lesson, love, and seized my heart and from a fragment cut keys to the storehouse of reality"!

Another famous woman mystic, Rabiya Al Adawiya, intoxicated with love for God, wrote, among renowned poems, this love intensified sonnet.

"O God, if I worship thee in
Fear of hell, burn me in hell;
And if I worship thee in hope of paradise,
exclude me from paradise;
But if I worship thee for thy own sweet

love's sake, withhold not thy everlasting love's beauty!"

The famed Persian mystic poet, Jalaluddin Rumi extols divine love in his ode,

> "O my soul, I searched thee from end to end;
> I saw in thee naught but the beloved; call me not infidel,
> O my soul, if I say that thou thyself art he."

There is a vast corpus of famous Persian and Arabic mystic poets, celebrating God, the one Supernal Beloved, to even begin to include in my so brief space. One other last brief quatrain of Ibn Arabi, expresses God's indivisible union in man and the Islamic revelatory light of Truth/God that images through man's spiritual eye as one in vision and unison.

> "When my Beloved appears,
> With what eye do I see him?
> With His eye, not with mine,
> For none sees Him except Himself."

Although the lot of most humans rarely rises to deific ecstatic levels of such inspiration, exultation and joy, love/God none-the-less is imperative to spiritual, physical embodiment of optimal health, and immune equilibrium, for ultimately Love heals. Love spiritually uplifts, recharges and revitalizes the entire human immune system!

The worlds healthiest people, the Muslim Hunzas, pre-eminently epitomize and incorporate the spiritual veracity and corporeal physicality of peer-less health, exuberance and jubilance, wholly disease free and life expectancy of 150-200 years! What more proof does one need of God/Love's pre-eminence and viable living in oneness with the creator beneficence! Prayerful repartee x-times daily and serene unison of mind, heart and soul, plus nature's idyllic diet regime of organic plant proteins and whole grains subsidize optimal perfect health for man as a non-carnivore species. We, humans are shaped, as our primate ape cousins with miles of intestines, totally unsuitable and impervious by physical nature 10 flesh consuming carnivores like the cat and canine dogs, etc. Although we do consume red meat, our digestive systems can, in no way tolerate constant consumption, except with dire disease consequences, conspicuously cancer and heart stroke, and vital valves circulatory blood flow stymied and stricken.

The Hunzas have no such illnesses, disorders, to reiterate, totally "disease free". Besides all vegetarian organic green plant food, live fruits and raw nuts (highest plant food protein), Hunzas consume the Semite Muslim "whole wheat" nature's perfect food, combining complete vitamins, minerals and trace elements that protect the human immune system. Yoghurt and raw honey's intake purify and energize the entire neuron physical enterprise. The proof is realistic, vis a vis all other life styles and diets,

Asian and western. Mankind falls grievously under our maximum health and longevity nature designed, and engineered for our homosapiens species! Let us learn the "real" spiritual and physical perfect health, and exuberance of well-being from our Hunzas brethren. For the Hunzas are earthlings like ourselves, not from some distant planet, but of humankind, such as us!

The Hunzas unconsciously practice Islam's whole soul, mind and bodily science: with the exception of diet regime, which for the vast Muslim nations, does include lamb and beef, though organically consumed and kosher slaughtered, according to the Semite custom carried forth by Islam. No animals, that suffer, or die of their own accord are eaten-again to preserve human health overall. All meat is consumed with the addition of "bulgur wheat", which yields fibre for easier digestion and lessens red meat quantities consumed. Pig, or pork is strictly forbidden, in the Semite tradition, as Islamic scientists found that pigs are carriers of toxic bacterial viruses, hazardous to man's metabolic immune system. The dreaded "swine flu", which tragically killed some 50 million European and American victims, following the first world war, never touched or afflicted, in the least the vast Muslim communities, nor orthodox practicing Jewish peoples. Live organic greens and vegetables form the greater part of the Islamic "Hal" organic diet. And we westerners have learnt, as of late, to enjoy the raw salads of "tabooli", raw parsley, scallions, and bulgur wheat added, together with "humus", garbanzo beans, sesame oils and garlic. "Sesame", an Arab word, literally means "open up vitality"; discovered by the Islamic scientists to release vital enzymes, multi-minerals, trace elements, and expressly increased the heart's stamina, and preempted sexual vigor and vitality. Islamic foods cuisine has preserved the "bible foods" tradition of the Levant, or Lebanese inheritance! Scrumptiously delicious and optimally health fostering. Raw garlic constitutes nature's most powerful anti-biotic: which excites the circulatory blood flow, thwarting the accumulation of fatty cholesterol deposits, protecting the heart valves functions and detoxifying the other vital organs of liver, kidneys and the entire human immune system.

As Islamic science all-encompasses the world's first comprehensive ecology of universal nature, in both its esoteric spiritual dimensions and exoteric physical realities: the honeybee, and its marvelous performance and produce, amazed their biologists. Moreover, the holy Qur'an contains a whole chapter revealing the bee's distinctive role- for the superlative benefit and blessings of optimal health conferred on mankind. The honey-pollinating bee is the only animal insect that can be domesticated by man. And honey's nectar sweetness irresistibly appeals for nature's (albeit amusingly overt) end results - of multi-endless benefits to man's immune system balance. This "divine golden nectar", inverts to boundless vital energies, powerful antibiotics, killing germs, bacterial toxins and viruses, and by its larvae trace elements fortifies and regenerates man's entire immune system! Raw honey is the only sugar that does not have to be converted into sucrose by the body's metabolism; and in Islamic medical practice, newborns are fed honey, besides mother's milk. Diabetic afflicted persons can easily consume honey, excepting our modern physicians appalling ignorance, or stubborn refusal to subscribe to any method other than their drug confined, conformed consensus. The bacteria resistant honeycomb composite does not decompose or deteriorate, like all other substantial matter. Honey was found intact and edible, after 6,000 years in the pharaohs' tombs!

Lastly, we reiterate that the Islamic scientists, upon intensive examination and research of the bee hive construction, mimicked and adopted the bee hives' intricate construct colossus, and applied the bees monumental marvel to Islamic architecture, creating arches, successive archways, and towering vaults, without the supporting pillars, used by the ancients, heretofore Greeks, Romans and Byzantines, Cordova, Spain's' magnificent Al-Hambra and other Muslim edifices attest to Islamic architecture in its illustrious grandeur of heaven-bound beauty and intricacy! Later, the principles and building codes of Islamic architecture was used by Europe, in their glorious towering cathedrals.

To return to ascribing "Unani" medical science, and attaining to near perfect health: e.g., firstly eliminating all alcohol and drug intake, as reverberated over and again! We can also opt for absolute perfect health disease free, and 150-200 years longevity: if we adopt the Hunza non-carnivore diet, distinctly designed by universal nature for mankind!! The great ape, namely, the orange orangutan, comes nearest man's stomach and intestinal digestive functions, and consumes raw fruit, vegetables and nuts, living up to 150 years. Man is favored to outlive our ape cousins by another 50 plus years! If we only hearken to our own select physical potential disease free longevity!

None of the world's healthiest fortifying foods of Islamic organic agricultural development would have been possible or yet available, without Islam's monumental horticulture advancements, and the art and science of Islam's vast geological researches and datum findings; faithful and at all times in accordance with nature's beneficent wisdom, in providing mankind's optimal food nutrition!

Water (H2O), notwithstanding its spiritual components, is mentioned in the holy Qur'an, as the primary substance, from which man is formed; plus purifying and exhilarating enlivens man's composite being. Water, is a magical substance, conceived for man and all living being. The "Hama'am" bath hygiene was pursuantly introduced, was, and is still held paramount to sustaining physical health and vigor, and detoxifying by hot steam mineral baths, which were daily habitual, even without modern plumbing amenities for effectually and beneficially ridding toxic poisons through the skin, tissues and muscles. These Hama'am hot steam baths later became famous as "Turkish steam baths", later adopted by European Scandinavian and health proponents.

The dreaded AIDS and HIV diseases are non-existent among Muslim-kind, anywhere within the vast Muslim populace, (which confounds the United Nations' Secretary-General's consensus). Why a strange world phenomena? Only because the holy sanctity of marriage between man and woman prevents sexual contaminant viruses and precludes what Islam deems immoral perversity. Counter, and in violation of nature's inviolable laws of male-female electro chemical balance! And human immune equilibrium, plus nature's designed mother/father governance protects and preserves the birthed offspring and their highest well-being! Lest we forget the sexual vibrancy and stamina that maximum health rewards. The Muslim Hunzas, mankind's healthiest specimen of both man and woman enjoy sexual vibrant vigor and stamina rigor, extending far beyond western spouses. Women's "change of life" occurs after 75 years, as she mothers newborns from 65-70 years, and men father offspring well over 100-125 years of age. Marriage is faithfully monogam-ous, while divorce is permitted by right of both sexes, love, overall rallies nurtures and harmonizes God's blessed bliss and faith-abiding family unity and unification!

Again, we must reiterate the pre-eminence of our spiritual veracity, through prayer's ascendency and liaison repartee and unity in our soul "Ruah", infinity, all embracing ineffable oneness "Al Ruah"/God; and love's unison of joyful being in the all infinite being, "Al Kindi"/God, from whom ignites this eternal "elixir" of life regenerating, revitalizing spiritual vivacity, which fosters exuberance in all health, love, jubilance and pure elative peace! We have over and again resonated the stark reality - that mental illnesses, depression, and Cancer are virtually unknown among Muslim-kind, unless they have adopted the modern secular perspectives and denaturized junk food diets of Westerners.

We of western humanity can rediscover and recapture our own God-endowed spiritual "elixir" evolving optimal peerless health, boundless exuberance; disease free, no organ implants or body parts and applicable surgery, needed; jubilant life expectancy of 150-200 years, if we subscribe to the Muslim Hunza's perfect non-carnivore, organic raw live plant, vegetarian, fruit and nuts protein diet; as Universal Nature Creator, "Al Fitrah"/God elected for his man created humankind!

CHAPTER 5

"HOMOSEXUALITY" - HAZARDOUS TO YOUR HEALTH"

The overall reality must at first outset be apprehensible by all adherents to our western Judeo-Christian American civilization: that our secularist populist culture of the "gay rights" issue and "same sex marriage" legitimatized by the United States Supreme Court this epic year, 2017: unequivocally imposed over all opposing dissent - be it "Christian", moral ethos or otherwise contested - as by the "non-polled" majority of American citizenry; succumbed now to the new profane politico secularity versus sacro-God sanctity. We, who cherish those God ascribed principles of moral veracity over secularist humanist proscriptions, of fibre core Christian values, established and enshrined by our founding fathers underlying truths, vouchsafed by nature's inviolable moral laws constituting humankind's highest potentiality for perfecting our species God endowed propensities and endeavors. Now, only to witness those sacred absolutes of truth vis a vis profane relatives; and decry the obvious denunciation and degradation of our American civilization. However the yea-sayer, politico populists plead or justify their "same sex gender equality" under man fabricated laws, let us identify our American state of affairs, apropos its name and stark human degenerate condition: profane secularity personified? Devised outside the realm of God and the absolutes of truth... Call it what you will; nothing less than extremist egregious secularism; in no way "secular" as defined by American democracy; but a contaminating "secular scourge" that defiles the holy union of man and woman-man concocted, devoid of ascendant God ascribing options. Secular, sodomist, sexually demoralized desecration. Godless "egolatry": worship of ego over God. Man's own make-shift fallacies over truth's absolutes; non-conforming to man's superficial enterprises and devious deceptions! The vacuous argument for "gay rights" acceptance, that God loves us all, albeit heterosexual, man, woman or "homosexual, lesbian or transgender is unequivocally true! But mankind is still subject to nature's inviolable laws, and those who violate those irrepressible laws, suffer the dire consequences of their own self-inflicted harm, albeit their health, mental, equilibrium and peace. Man cannot contrive his or her own willful maneuver, can never outwit, out-step or outmaneuver universal nature's implacable laws decreed for humankind. Our will, inevitably must align with the almighty's will. "Thy will be done on earth as it is in heaven," expressly applicable to the universal condition of all human earthlings. Only the God Supernal Creator of his own man creation- be uniquely the all omniscient, all-knowing, all infinitely wise of his man species, and the blessed pair of man, woman and procreating off-spring. However we "pre-fabricate" and distort, dispute, and disclaim the infinite omniscient God creator's elect will and pre-conceived blessed purpose for mankind!

We, who cherish our nation's founding religio-spiritual God-inspired tenets of preservation and protection for all mankind, lament the demoralized degradation of our American civilization that has legalized abortion. To mercilessly destroy a human being which science attests is formed at conception, 23 chromosomes each from father and mother. The relentless abortion advocates' insidious guise under "a woman's reproductive rights" translates "the right to kill" an unborn defenseless babe, already pulsating, living within his or her mother's womb. We foresee how future civilized human societies will judge American aborted millions of human lives! Abortion is ranked as heinous crimes against humanity's precious vulnerable defenseless babes beings. We will cover more at length, the abortion carnage in my chapter on womankind.

Space is too limited to divulge the other horrific acts and crimes against American innocence, like a beautiful wide-eye "baby soul" discarded on a beach, in a garbage disposal bag, or touch on the heinous crimes of incest and sexual molestation and rape perpetrated on a daily basis. Sex trafficking lucrative prostitution of children and minors, to satisfy depraved perverts. America's wanton senseless street crimes, violence and relentless deaths beset unwary innocent helpless victims. Less than a century ago, our homes, families, neighbors and communities were safe refuge and moral conscionable bastions of

American brethren!

Regardless of those of us who gravely differ with our American nation's political referendum and moral agenda, this chapter on homosexuality, is in no way intended to besmirch, revile, or least of all, condemn our lesbian sisters and gay brothers steeped in same sex behavior. But I pray will solely serve to manifest the deadly health hazards and tragic lethal untimely deaths inflicted on same sex life styles.

Our medical doctors, intimidated by our country's powerful political conformist gay advocates lobby have forfeited their solemn dedication to medicinal humanitarian service; failing to disclose and forewarn the life-threatening disastrous consequences of engaging in same-sex lifestyle and practice. So entrenched has "political correctness" and political liability for "gay orientation" seized conscionable medical science's commitment to safeguarding and optimizing human health!

Auto-Immune Disease Syndrome (AIDS) now constitutes humankind's deadliest and devastating, degenerating disease, worse than cancer, which offers partial or survivor cures; whereas aids has no cures whatsoever - only total abstinence. Promising treatments and ever-new drugs comprise cruel lies and false hopes for AIDS and HIV-infected victims. As AIDS (Auto-Immune Disease Syndrome) certifies, the human immune system becomes disrupted, disoriented and disabled. The wonderful protective immune resistors now fail, and the entire human metabolic organism falls vulnerable to invasive viruses and toxic contaminants. Few of lay patients, eager to assist in any way, know that blood transfusions from AIDS and HIV donors, are outright refused by physician practitioners and the Red Cross blood bank, to bank, to ensure blood transfusions, safe for all patients.

The "gay rights", however their advocates have pleaded their case, has never been a "civil rights" issue, a distinct misnomer, vis a vis ethnic and racial basic human rights, which signify real human "civil rights". Clever deception, never fool the common sense intelligence of the bulk of Americans, albeit unawares or perchance ignorant of the gravity of self-same sex repercussions on America's future posterity and diametric health hazards. The ultra liberal political machine has unwittingly dis-serviced their unassuming constituents, in winning the popular electorate vote, notwithstanding the detrimental demoralizing debacle, in de-stabilizing the sanctity of family and community unification! And, as reverberated, in disregarding and desecrating the pinnacle sanction of holy conjugal union and marriage between man and woman, and their natural birthed offspring!

Universal nature's inviolable laws govern unalterable and immutable, nor compromised by man's superficial suppositions or divisive digressions, via same-sex hypotheses. Nature's vast living sea and terrestrial eco-system's vegetative plant, animal and mammal being are governed by unalterable male and female magnetic unity; that opposite sex attracts and like sex repels. Nature's procreative process, propagates the survival of each species. Thus, this universal eco-chemical balance of plus and minus opposites operates irrevocable; however the man mammal contrives his or her sexual preference or perversion for "like", or same sex performance. Such who engage in violation of universal nature's immutable laws... And willfully digress, suffer alienation from their own God dispensed nature, with dire consequences of mental, physical dysfunction deterioration, and degeneration of their vital immune system.

Thence, the inevitable deadly AIDS and HIV life-threatening diseases, disasters, and foregone tragedies.

We have divulged elsewhere, that Islamic physicist-chemists had first discovered the electro-magnetic chemical opposites of positive, "plus" and negative, "minus ", or electrons and protons which opposite lobes construct, harnessed electric generators, that lit north African cities, 1100 A.D., centuries before western European and American scientific successes.

These same unalterable and irreversible laws of universal nature, that opposites magnetic positive and negative govern all plant, insect, fowl, fish and animals unifying propagation and survival of their distinct species; impacts man and woman's complementary chemical balance, albeit human nature rewards higher spiritual elative and aesthetic awe-climatic inspiration! "Viva la difference," O joy of joys prescribed by nature herself!

Besides the truly loved and cherished spouse conjugal joy, marital bliss, companionship and blessed parents' bond: there does indeed, exist nature's God inspired pure platonic love-that binds two souls, of either gender, between sister, brother or as sister and brother, too often mistaken by outsiders, as lesbian or homo-sexual. This more pure, spiritual souls' radiance-may even prove more powerful than all other loves; inspiring intellectual illumined truths, entrusted companion, champion and confident in perpetuity, who outweigh and outlast wedlock, family and friends of the flesh! History has celebrated such "soul mate friends", be they ever so few and far between life's encounters.

We have reiterated that AIDS and HIV diseases are unknown and non-existent among the Muslim populace throughout the globe; a startling fact and reality that astounds the United Nations pre-eminent Surgeon General. There is no mystery, I have explained. Islam, above all honors the conjugal holy union of man and woman, as we Christians, once upon a sacred time, also sanctified. And, we should as well communicate Islamic culture that recognizes the pure, undefiled brotherhood of man to man, as "brother", woman to woman as "sister", and man, woman, as "brother" and "sister"; sans any sexual implications whatsoever, homosexual and lesbian characterizations are considered and deemed perverse and depraved, unequivocally adversely in violation of God's natural inviolable laws. The holy al Qur'an confirms the gospel's sin of sodomy, and the west's behavioral same-sex issues have never posed problems for the world's most God-oriented, God-centered, God-sanctified human society on the face of the earth!

Since it is universally acknowledged that both man and woman possess each other's gender opposite masculine and feminine characteristics and qualities: identified in Asiatic science as "Yin" or "Yang"; lamentably, in our western present day of political "same sex demagoguery", these sexual cross-contingent inherent male/female attributes are misconstrued as "gender identity", and tragically, to the detriment of nature's complementary evolvement of the whole integrated human being. Woman's, sometimes, over-emotional propensities, need a man's inherent more aggressive sustaining strengths, and man, woman's more tender sensibilities and mercies thus, nature endows both, with the spiritual and mental facilities to confront life's unforeseen problematic circumstances, and adverse traumas.

Beware, parents, as true guardians of your precious children: be not persuaded and deceived by the prevailing popular gay-lesbian or trans-gender hypothesis; that your boy or girl child possesses sex genetic flaws. Universal nature is flawless perfection, apportioning complementary attributes for sexual eco-balance. Only man's abysmal flawed rationale and ignorance of nature's supernal intelligence and inimitable wisdom, succumbs to his own heedless crises and deplorable, defeasible high-spirited, rambunctious girl, or "tom boy", or tender, sweet sensitive "lover boy", merely reflect his or her complementary integrated whole being. Let a girl play soccer, football, or swing a bat, climb and dangle, "hang-out" in trees; or a boy child masquerade in his mother's shoes, hair coiffure, or dress fashion. These are normal, natural behaviours, that in no wise, configure gay, lesbian or transgender tendencies, but God's preconceived designs for the whole integrated compliment of man and woman being! Beware, my fellow mortal peers - that you do not "play God", the one and unique un-creatable immortal Sovereign Creator, who flawlessly creates the likes of you and me!

The Eternal and Almighty God, from inflicting gravest danger to our innocent children, with falsifying disconcerting mental, neurological-spiritual populist suppositions that breed rank confusion and depression, resulting in the break-down of their healthy body's immune defences and equilibrium balance!

Save us from misleading our innocent pure babes from believing his sexual identity be far removed from his nature's inborn beauty, reality and invincible truth. That we, as God bestowed parents as entrusted guardians, fierce protectors and sagacious seers, guide our precious innocent ones, so wholly dependent on our wise guides and integrity of holy truth!

As long as our gay and lesbian brethren persist in their operative life styles, they incur the devastating health hazards and ravaging AIDS and HIV diseases, with catastrophic immune fatality and untimely tragic death. We need our physician doctors of real science heed the incontestable truth of homosexual carnage of precious human lives, and warn head-on the inevitable doom.

We are in dire need, for a moral, spiritual resurgence and rebirth of our American democracy, God-inspired and God sustained. We must needs breed a new generation of great genius intellects like Ralph Waldo Emerson, Will Rogers, Mark Twain, Walt Whitman and Emily Dickenson, et cetera, devotees of immutable truth over ephemeral falsehood. Tenacity of moral veracity over vile... defiled depravity... sacred God pre-eminence over profane vulgarity.

We need untainted, incorruptible political moral leaders, like Abraham Lincoln, Woodrow Wilson, Theodore, Franklin and Eleanor Roosevelt, John Kennedy and Ronald Reagan to lead us out of our demoralized, degradation of religio-moral Christian values and vitality! We need to retrace and revisit our God ascribed principles of moral veracious resiliency that created the greatest nation on earth, the United States of America! God bless and re-bless America, and as a God beholden beacon for the entire world!

CHAPTER 6

"WOMAN-KIND"

Woman is not equal to man, nor is man equal to woman: except in their compliment; as one, in unison with each other. Women-kind, since humankind's advent, whatever the clime, epoch or time, has proven man's integral, exigent complementary ally, companion, lover, mother, spouse and sustaining moral stamina... Endowed by nature with emotive love's gentler sensibilities, her irresistible tender loving powers have won her mate, overwhelming supremacy of reign in the sanctity of domicile refuge and family. Woman has proven man's moral superior, by virtue of her inborn fierce maternal instinct for the preservation of her own and mate's offspring and survival of the man species. And as such, a priori, womankind serves as the explicit "governor of mankind"

Modernity's western "women's lib", for "equality rights", belie the reality, that woman's singular feminine powers of gentler persuasion and man, woman complementary unity, have already won her man's inspired allegiance and counterpart diversity ad infinitive perspectives and intelligence! Sadly, too many women turn abrasive and lose their coordinative and cooperative complementary alliance to antagonistic competitive militant aggression, in futile frustration and mistaken tactics, pleading unequal treatment and gender injustice; when these so called victims, royally play ignoramus! If any woman utilizes her own inherent unique gifts, genial capabilities, genius of kind, no one and nothing can hinder or deter her limitless success! It's written in her own stars! Our world has produced outstanding, distinguished women in every field of endeavor; no excuses or intervening unfair maneuvers ever deterred their accomplishments! Self-reliance was forever the ace individual persona motivation and persistence for wondrous achievements, come what may! I need not enunciate renowned womankind, herein this so limited space, but the literate among us, are well apprised.

Needless to say, innumerable western renowned men have paid great homage to their mothers, spouses and other noteworthy women, who impacted their lives; whose successes and triumphs were inspired and championed by their women folk. Mothers, who weaned their sons as moral voracious leaders, and devoted spouses, who impacted their husband's upsurge careers, inclusive of America's famous political statesmen, who guided the public weal for justice and reform, and illustrious transformation! Such mothers, spouses, and family patriarchs inspired human history's greatest advancements; in our century, the Roosevelts all - Teddy, Franklin and Eleanor, recognized America's disenfranchised and disparate inequalities as grave injustices. Among other civil beneficial acts for the betterment of the common man, "social security" was enacted to secure economic guarantor after a lifetime's arduous work and toil, for the future elders unforeseen survival needs.

We have reiterated the disparaging, demoralizing degradation of our once proud American nation-commemorating a United States of American democracy, without monarch or hierarchal oligarchy; but only to God ascribed moral efficacious priority of principles and civil, judicial equanimity for common-place conscionable man. Decency and upright decorum and behavior comprised common codes of ethics and mores. What was once considered sexually perverse and "sodomist sex", is now seen as permissive, licensed law of this once God-fearing, abiding land. But the higher law of Divine sanctity supersedes all man-fabricated laws; and universal nature's inviolable laws of unwavering, irrepressible morality, ultimately exacts and redeems. None can transgress nature's immutable laws of good versus evil – superficial for Immutable Truth, however mutable man contrives!

What was hallowed the sanctity of life, and the new unborn babe in the mother's womb, is now defiled and destined for slaughter and abortion infanticide, by the successful advocates of "a woman's right to choose", or rights over her reproductive properties. Yes indeed, the "right of choice" to abort and murder her unborn child, women, as mothers, above all other beings - God bestowed holy entrusted stewardship, to ensure the survival of our human species. However they justify their crime of "womb babe infanticide", they must render to a higher Law of moral retribution, than the United States' law of abortion. These same abortion advocates are more zealous in saving the endangered polar bear, than their unborn child, defenseless, vulnerable to the merciless extinction of his or her unborn life! The American, unborn human beings are veritably the most endangered of all species! Pray tell, what if these ruthless women had been aborted by their own mothers! Never to experience the marvel of being! Unrepentant, ruthless selfishness knows no end!

"The moving finger writes, and all thy tears and prayers can in no wise, wash out one syllable but the Supreme Overseer knows all!"

The American nation has depopulated by millions, due to the abortion laws, by extinguishing millions of human unborn lives, together with Europe and Asia, and the entire world, except Muslim-kind, who still sanctify all living and unborn humanity. We need to address these two, so vastly conflicting, contrasting societal cultural divergent norms, betwixt western women and their eastern Muslim sisters. The Holy Prophet of Islam laid the "first norm" of mother's exalted station in Muslim civil society: "Under the feet of mothers, doth heaven lie!" Womanhood is prized as man's moral superior, recognizing a woman mother's fierce protective instinct for the preservation of her unborn and born and woman's key formidable role in the survival of the human species at large. For the holy prophet revealed all humanity, Muslims, Jews, Christians, non-Muslims and non-believers, "all embody one human family"! Islam is ever focused on mankind's higher evolution and perfecting evolvement, "the perfect man", "Al Kamal Al Insa'an". Ever mindful that El Mohammed was the posthumous "Christ counselor of truth" to rectify the Christ messiah's mission. "Be ye perfect, even as your father in heaven is perfect"! Jesus, the Christ messiah had revealed man's divine link to our heavenly father. Islam verified mankind's God-forged divine implant in man's "God inherent, innate nature" - "Al Fitrah"/ God (Sura 30:30) - the heart of the Islamic revelation. Thence, Islamic moral veracity must align all phases and conditions of life. As distinctly applies to both women and men. How could the thought ever enter the mind of a Muslim woman, of aborting her unborn child, when Islamic precepts were practiced on a daily basis, and prays x-times each day, with heart and mind God-ward, elevated the everyday mundane! We have reiterated, that sans church, mosque or institutional religion, sole authority was, and is delegated to God only, without man surrogate interference or intervention. God's Omniscience supersedes all others, within full freedom of single, individual conscience. The very word "secular" does not even exist in the Arabic language; for as all languages and entire cultures manifest their underlying psychic ideologies: Islam identifies God, as the Infinite All, the Omnipresent here and now, within the inextricable living fabric of human

116

existence. No secularist divergence or invasive alter-social, political populist contrivance intercepts God's supremacy, as the Infinite One and All, permeating all universal nature, and the underlying Immutable Truth and Reality. ("Al Haqq"/ God)

Not plagued by western secular populist prototypes of gender inequality: our eastern sister's truer womanly purer, loftier, near-ideal sensibilities, come to the fore, in uplifting and elating her mate; her tender sensitivities caress her lover-spouse's baser bestial instincts, console, comfort and calm his more aggressive proclivities; while her intuition of intrinsic wisdom intrigue and mystify him and all mankind! She remains a paragon of constant amazement, mesmerize and surprise! Uppermost, the true woman's ideal, spiritual regality and pure aesthetic beauty captivate her mate's baser physical opposite gravity. Arabic and Persian poetry laud and celebrate a woman's love-hypnotic spell, her elative grace and illustrious beauty, above all else, her eyes soul shimmering stars, more dazzling, that shame the celestial stars! Endless love poems and songs enhance and exalt womankind's beauty, by any other comparative measure of past or extant literature! Sex, is not "sex" per say, but always linked to the holy beatitude of love first and eminent, as God is ever fervently praised, as blessed Bestowed of man and woman's beloving unity! Thus, doth God, give holy sanctity to spiritual, mental and sexual physical joys and pure rapture. The holy Qur'an warns of mere physical lust, in favor of spiritual all-embodying pure love! True love encompasses all, embracing heart, intellect, consort, companion, sustainer and enveloping consoler and champion! Nowhere is woman-kind more cherished, extolled and celebrated, either in western or Asiatic culture and documented literature.

Woman's highest acclaimed sublime station, veneration, and ethereal adoration is epitomized by the "Taj Mahal", the Seventh Wonder of the world. Whose sole love was eternalize by one shah Jahan's beloved tribute to his adored spouse, soul-mate, constant companion and co-ruler of twenty-years -"Mumtaz". Shah Jahan relied on his cherished intellectual and moral "superior" as pre-eminent participant governor in his everyday judicial rule of his Indian Islamic empire. The architectural splendor of the Taj Mahal deliberately intimates and foresees the holy Qur'an's revelation of paradise, "Jana's God-wards transcending beyond earth's confines. Never more such a replica of heaven has yet to be symbolized... "In my father's house, are many mansions" and when we humans grasp heaven's parameter in the after life... Al Qur'an reveals: "Where have I seen this wondrous beauty before, recalling earth's splendor!" The author is preparing a film script on this, the world's greatest love story.

Since God's surname is "Beauty" - "Al Jamil": who creates woman's wondrous beauty for her equally man marvel of masculinity... Together with all universal sheer splendor ... And magnificence earth aghast with awe-wonder, pristine wilderness, soaring eagles, birds on the wing... Ravishing flower blooms; and New Hampshire's impassioned autumn glowing crimsons, scarlet and gold! Dare I reveal the obvious truth... that God's manifesting beauty is "romantic"... That God is the "Ultimate Romantic", from whom all infinite beauteous sources spring! For "Beauty"... "Al Jamil"/God creates, shapes and recreates all universal living things and being! And "Al Jamil", "God, Beauty", doth indeed create all living phenomenon being, expressly in loving forms and "Love/God" "Al Rahman"... Forged and shaped in love's image! God verily loves what he creates! Ibn Arabi resonates: "Were it not for love, (and love's inviolate beauty) God would not be worshiped!"

We have endeavored to unfold the exalted status of womankind in Islam, still prevailing among international Muslims' religio-societal, cultural dynamic. The historical reality of Islamic civilization over one millennium, founded, formulated, established and promulgated universal human society's first civil and judicial laws of woman's equality, justice, liberation education and ethical supremacy, barely established in our American democracy, less than a century ago. Islam, above all the world's religions, elevated womankind, and emancipated women, as men's equals. Women were conferred sacred and inviolable rights of protection, financial and moral support, education and full participation in Islamic society - inclusive of non-Muslims, Jews,

Christians and minorities of all ethnic and racial diversity. As we have reiterated, Islam never suffered our western racial disparities and discriminations... the world's first successful egalitarian brotherhood "Umma" and still perpetual. Womankind won individual rights, in retaining her own family inheritance, and deeded property rights, and equal custody of her children, in the event of divorce (while rare). Women retained their maiden names, in marriage, not their husbands, a practice adopted in Muslim Spain, and still distinguished by Spaniard women today.

The magnitude of historical, documented Islamic law of women's total liberation and unstinted equality appears to have escaped the present-day terrorists, Al Qaeda, Taliban and ISIS, either wholly ignorant of Islamic history as well as the holy Qur'an, whose divine revelation endowed all womankind with full equality with men, posthumously establishing Islamic laws that safeguarded women's' sacred rights. These demonical terrorists' non-Islamic barbaric, brutal treatment of women, of any class, religion or ethnicity: defies all "Shariya" Islamic law towards nothing less than reverence and full sanctity of protection for all womankind! Damnation, oh bedeviled terrorists, hell-fire is promised you, for your heinous crimes against women humanity! In the holiest name of "Al Islam", "peace" and infinite compassion and mercy "Al Rahman al Rahim"... Almighty God's pre-eminent name and invocation, for all mankind to follow, and true Muslims leading the path, "Shariya" God-wards in unifying love, peace and equanimity for "all the family of Adam and mankind"!

Education, scientific enlightenment and the uppermost endless quest for all universal knowledge, and immutable truth "from the cradle to the grave", is ever incumbent on womankind the principal tenet of al Qur'an. Thence, as we have reverberated: Islam ignited the world's monumental scientific discoveries, revolutionary, evolutionary and underlying universal truths which, ultimately saved medieval Europe's dark ages and gross superstitions, and abysmal ignorance. God's pre-eminent holy command: to seek all knowledge and "scientia" as sacred and uppermost, blessed, uplifted and transcended darkling Europe to renaissance enlightenment! For God, the Infinite Omniscient "The Unseen, hidden Immutable Truth and only Reality. "AL Haqq", has willed all mankind know the light of universal truth! For God elects his human, "the knowledgeable creature" "Al Insa'n" as God's contemplative companion, imaging and mirroring God's lightening truth! Not Eastern, Islamic, nor Western, Christian, but encompassing all global humankind!

Islam produced innumerable famous women poets, philosophers, mystics, scholars, and the world's first doctor physician, most of whom, have never been translated into the European Anglican languages. Such renowned women, sat in the world's first university-mosques with men devoutly assembled as student audiences; more attentive to womankind's intuitive, inciting gleaning mind, than their own male scholars. This script falls too far brief to divulge the great echelon of illustrious women; who also inspired other Arabic and Persian Islamic Greats, who followed their lead. Rabiya Al Adawiya was one of' the first mystic-poets, whose spiritual ascent God-wards, in adoration of pure rapture, gave audacious transcendent expression and pen, to a massive Arabic Iranian compendium of the world's greatest mystical poetry, ever articulated anytime, anywhere!

We quote a few lines from the Arabic famous woman mystic.

"Oh God, if I worship thee for fear of hell, burn me in hell! And if I worship thee for hope of paradise, deny me paradise! But, if I adore thee, for thy own Love's sweet sake, withhold not Thine Love's everlasting beauty else I perish in the forever now!"

The God/love ecstatic Rabiya further enunciates...

"Two ways I love Thee, selfishly, and next, as worthy of Thee.

'Tis selfish that I do naught, but think on Thee with every thought.

'Tis purest love, when Thou dost raise Thy unseen Veil, to my   adoring gaze.

Not mine the praise... But Thine is the praise in both Thee and me."

"Ye must die, before ye die", utters the Prophet of Islam. Die from self/ego unto God... The Infinite, Ultimate One Reality. ("Al Haqq") Later, mystic Muslims called "Sufis" para-phrased the Holiest Prophet's revelatory tenet... "Ye must die, before ye die"... As "fana", "passing away from self" while the Sufis devised an entire system of "steps leading to the proximity of the prophet's exemplary nearness to God, and the "Intimacy of all intimacies": the Sufis missed the Holy Prophet's revelation of man's immediate access and available ineffable unity and oneness in God, without any instructive or doctrinal steps or procedures!

Islam is without embellishment, indoctrination or dogma: for to only reiterate again, the heart of Islam: that "man is forged with God's innate nature", "Al Fitrah" (Sura 30-30) He, man, has only to connect in unity, with his God-endowed, inherent nature.

The following verse expounds the prophet of Islam's revelatory "dying from self", and herein the adoring Muslim mystic's misery, in realizing once again, his ecstatic union with God, the One Beloved. Al Nuri pens the following, although some western Asians attribute the lines to Al-Junayd.

"I had supposed Thee, having passed away

from self in concentration, I should blaze a path to Thee. Ah, but no creature may draw nigh Thee, save on thy appointed ways."

"Lo! I have severed every thought from me, and died from selfhood, that I might be Thine. How long, my heart's beloved I am spent. I can no more endure this banishment!"

Saints and mystics were linked directly to God's overwhelming, overflowing love, with no doctrinal interceding or hierarchal intermediary only God, the One Supreme Authority, the one and unique Beloved... "Whose throne is the heart of man?" The medieval church, with its hierarchy and authoritative control, discouraged the free, direct affinity with God, as enjoyed by Muslim mystics.

Saint Theresa of Avila, Spain, enamoured with Rabiya Al Adawiya and other Muslim poet-saints, created poetry and elative verse, nearly duplicating the woman mystic Rabiya and other Muslim mystics. Not by "salvation", vis a vis Christian doctrine, for fear of hell and brimstone, but man is "saved", by virtue of his own God endowed, innate Nature, "Al Fitrah"/God. To reiterate, the Islamic heart of revelation, reveals the immutable truth of man's divine identifying nature, with all universal mankind. And man's God-inherent potential for perfection of homo sapiens. Thence-ward, prayer and contemplative repartee and impassioned aspirations for unity, love's exalted bliss and pure peace; must yet be proven wholly selfless - in loving service to God, through private and public acts that transform and edify the plight of humankind; the poor and needy, your neighbor, orphan and widow and all alike marginalised, be your brethren, all your family of man, to honor, uphold and enfold. These goodly acts bare greater testimony, to true love of God, and pure adoration-than all your endless prayers and supplications! Excellence of ethics, conduct and mores must supersede prayers in obeying God's pre-eminent ordinance of loving service to your brethren man, admonished the Holy Prophet of Islam. "Abdul'allah", "Servant of God" confers the distinguished appellation given one who lovingly serves the cause of humanity, and the epitome of Muslim faith. "The servant is greatest among you," Christ revealed, as he washed the feet of his disciples. To serve God is to serve man; love and service are one and the same.

This conscionable veracity of Islamic faith, performs today, in modernity, within the global context of humanity; with at least three Muslim women accorded the "Nobel Prize" for distinctive causes for human civil, scientific and artistic achievements, our Western critiques and judges, may hail their endeavors and accomplishments as purely "secular; not cognizant of the inbred Muslim dynamic for universal human social, cultural, judicial justice in transformation, and scientific and aesthetic, artistic edification. As we have reverberated Islam never suffered the fatal religious-scientific dichotomy of the West: God is all-infinitely pervasive in everyday life, civil society, science, and culture. God, is, infinitely fact and reality: the goal of all scientific and immutable truth and enlightenment.

Islam never decried, or suffered the west's God/science schism: of dead facts, soulless, without their Godhead divine idea; or chemical mechanisms, sans their spiritual identity. The Islamic science of "Tawhid" incorporates all spiritual realities, in perfect complementary unity with their material entities, indivisibly one, and whole-immutable. The Islamic visage of God's all permeating spiritual, immutable reality, encompassing all mutable materiality: has, as all science, pervades the commonality of everyday social and cultural mores dynamic; with both men and women; and since womankind comprise the foremost breeders, nourishers, sustainers and governors of mankind, the psychic, mind, and genetic DNA, passes on to their offspring's comprehension of all existence. Spirituality and God's conscious omnipresence is all-pervasive in common, everyday life. We have reiterated that Islamic societies comprise the utmost God-centered, God-oriented of human-kind, on the face of the earth. While the Western industrial nations, the least, and, in fact and reality, have inherited the science of Godlessness and rank atheism. Ego, has replaced God; and Western modernity has replaced the Christian faith, with the neo-religion of "egolatry", "worship of ego"!

While western science depicts a secularist, atheist Godless universe, sans a God Originator, Cause and Creator: Islamic science diametrically praises and celebrates the cosmos and man creation as a colossus of marvel, awe miracle and glory!

The family comprises the Muslim transmission of religious instruction and faith, over and above all institutions, mosques or schooling. And, moreover, since all external vehicles of Islamic faith fall short of Islam's quintessential, internal human fibre and heart of being. Reliance on one's God sources and resources within man's God inherent nature: "Al Fitrah"/God living all omnipresent within man. The Holy Intimacy is for each individual child to discover and grow awareness of him or herself, within his or her own God endowed singularity of being! Both parents are imperative to fostering their children's faith, stamina and resilience, in preparation to life's myriad of encounters and challenges.

As reiterated, the dreaded disease epidemic of AIDS and HIV infection are totally non-existent among the global world population of Muslim-kind; the elusive mystery, merely divulges the holy sanctity of marriage, as honored and practiced by humankind, since antiquity. No Godless secularist perversion of the conjugal unity of opposite sexes can, in no wise, defile or desecrate Muslim men and women's God blessed sanctioned unison. Unless, some "convert" to western permissive licentious sexual perverse ideology, and must necessarily confront the AIDS and HIV life-threatening disease and deathly calamitous end!

While western women clamour for Eastern Muslim women's "rights" and full-fledged freedom, except for the brutal crimes against women, perpetrated by the terrorist Taliban, Al Qaeda, ISIS, et al, invaders and oppressors of Muslim, non-Muslim, Christian and minorities' innocence; we have aforesaid, divulged the Islamic radical equality rendered women, in all facets of social, cultural judicial life, never before achieved by world historical moral codes and criteria; and the elevated transformation of women's status, revered, cherished, safeguarded and protected, for one millennium plus, within Islamic civilization, and continuing today within the Muslim international community, inherent and inextricable within the God sanctioned moral veracity of family and community life.

Contemporaneous Muslim women, enjoying such a sublime role, are not so eager to trade their lofty station, for the Western woman's freedom to freely encounter the stark reality of everyday predatory crimes of violence, rape and murder, so commonplace now in western, American everyday life. Freedom to traverse imminent endangerment, lurking in unsuspecting places, with no refuge for safe deliverance. The deadly secularist, Godless, heedless wanton crimes, are repulsed by Muslim women.

Modesty is Muslim womankind's attire for her own safety and protection. The burkha habit and black body-covering dress was never ordained by the Holy Prophet of Islam. It was historically introduced by the Ottoman Turks; it was in no way proclaimed Islamic by the ages preceding Turkish oppressive imposing rule. Arab women and all other Muslim women never veiled, or covered their heads. The Holy Prophet forewarned women, "to desist flouting your intimate charms for your own protection" true modesty guards against dangers of molestation and sexual assault. The present-day head and veil habit will inevitably pass and become an Ottoman tradition of the past, with the true Islamic social reforms needed to expel false Islamic beliefs and erroneous Muslim concepts, held and imposed by indoctrinated men over women. Guaranteed that Muslim women will never mimic the short skirts, body-tight pants and private parts indecency of Western women that sadly invite rape and crimes of sexual assault and murder. As alcohol and drugs are forbidden, no need to visit bars and cocktail lounges tragically, too many young American women have gone off, seduced and vanished, never to be found, or murdered. Albeit their parents' heartbroken loss, can never be atoned, the "freedom rights", without the necessary precautions safeguards for our vulnerable young women, exacts the ultimate cost for their innocent lives! The old-fashioned scoffed parental chaperoned youth, once practiced in the Christian, moral tradition, however outdated, aught to be revived as necessary safeguard protection for too many victims' disasters! Times, such as ours beg such safety measures for our innocence!

Our western American drug culture with alcohol, the worst addictive culprit; destroying lives and insurrectional wanton crimes of violence and countless innocence deaths; constitutes the major factor for the mental, physical deterioration, diseases, deaths and demoralizing de-gradation of our American Civilization! In reality, throughout the western hemisphere, drugs have ravaged scores of precious lives, and humanity, at large. We are overdue for a renewed prohibition of sorts, and/or the plain educational merits, to desist from liquor altogether. The monumental factor, as discovered by Islamic medical science: that "alcohol" ironically an Arabic discovery and word; prohibits alcohol intake, for the disastrous results of alcohol consumption, on immediate imbibe, destroys millions of brain cells neurons, and the neurological transmitters from brain to body. Muslim scientists and their corroborative physicians had unequivocally proven that heart vascular disease and cancerous interlopers are caused by alcohol usage, even in moderation. Alcohol was banned unconditionally, medically, judicially and within the cultural context of mankind's everyday life. It is highly noteworthy of significance, that both Coronary Heart Disease and Cancer are not prevalent among Muslims, the world over. That is, unless, they have taken on the western indulgence of alcohol. So how do we, my fellow Americans, solve and cure our nation's alcoholic habitual usage and addictive frenzy? For our health and immune system's better sake and stake?

Education does not stop in the Islamic milieu of learning and enlightenment vis a vis western academic higher university credentials. But incumbent and mandatory, "from the cradle to the grave": an endless process and quest for self-discovery and self-enlightenment, through "the university of the soul", and the cognitive search for the intuitive and ultimate Immutable Truth of Reality. ("Al Haqq") Of one's God innate identifying being. Inbred within the Muslim family, is the seriousness of this quest to know Immutable Truth or God - Al Haqq: for to "to know yourself, is to know God"; and "to know God, is to know yourself"; so intertwined is man's identifying persona within his God identifying inherent nature... "Al Fitrah"! All of Islam's religio-self-revelations emanate within the human enterprise, sans cleric or religious institution or human intermediary. Hence, again, the preponderant appeal to intonation-all mankind, as the world's consecutive fastest growing religion. And this, despite the non-missionary, non-proselytizing prohibition in Islam. "There is no compulsion in religion" one comes to God, of his or her free will and love/God's uplifting bountiful grace... "Al Baraq'allah".

Poetry, mystical and artistic creativity are commonplace among Muslim families youth, encouraged and challenged to express their own individual creative idiom, to explore their own unique creativity and self-discovery. We have quoted a few of the Arabic and Persian poetic Greats - who have been inspired within a God charged elative overflowing, impassioned love adulation of Love's Infinite Creator. Poetry, impromptu is a commonplace linguistic theme of expression, among the "lay" and poetic elite. "We are all poets," utter Arabian and Iranian common folk, not to be distinguished by Western poets of renowned Academia.

But family edifying, fostering the Arabic Islamic high nobility of mind, creativity and poetry to their offspring, has the constancy of lineage, fear of perilous destruction to family life and community safety; and touching every bordering Middle Eastern nation within the Islamic Arc; while the evil of Assad rains terror, genocidal war and destruction on his own Syrian people! Not since the Second World War has any human catastrophic carnage been committed. Even Hitler never warred on his own Germanic "super-race"; and the ovens for extermination, more benign than exploding barrel bombs and chemical gaseous warfare on devil Assad's own peoples, no town, village or city spared from his demonical ruthlessness! While the international community looks albeit on in horror, yet permits Assad's free, unabated license to kill, at will, humans, en masse, children, women and men, defenseless and innocent! And, my and your American nation, once the world's formidable moral leader, to implement America's own Christian core values of humanity; has forsaken that pre-eminent leadership, for political partisan conscript, and ignoble conformity. The now president, of these United States, who claims to be a Christian, no less, has utterly failed America's humane integrity of conscience and the world's, for moral integrity, and the stout uncompromising humanitarian leadership! Gone are the Roosevelts, Jefferson and Lincoln, who dared to challenge and rectify the evil wrongs, horrific injustices and inhumanities that afflict and persecute helpless, defenseless innocents!

When the body of a six-year-old Syrian boy washed ashore, the world was aghast with horror - sending shock waves around the world. He, his brother and mother had all met their drowning death, among millions of Syrians fleeing for their life, causing the greatest numbers of refugee exodus since Hitler's war! An amazing phenomenon, since victory and post-war international civil reforms and rectifying humanitarian transformations! And still, the new Hitler, Assad, is granted full dispensation, to continue to wreak havoc on innocent and defenseless civilians!!

Europe is already trying to address this horrific refugee catastrophe: while our U.S.A. has promised a fraction of the 12 million Syrian refugees; Germany's truly Christian Prime Minister, Angela Merkel, has welcomed over 100,000 Syrians, France, and Scandinavian countries, France and the United Kingdom taking in some minor quota.

Yea, we of womankind: endowed by nature, as love/God's elect primordial mother moralists, nourishers, governors, and vindicators of mankind, are ultimately the true saviours of humankind; our sons, spouses, neighbors, kindred brethren and the family of man. Men are born of women, and not the reverse, for nature's God definitive Divine purpose and edifying perfection of our human species. Woman is born a reformer, and celestial transformer, of the profane ephemeral now, to redeem our God inherent, innate nature and being. ("Al Fitrah"/God) We can and will vanquish war, aggression, poverty, inequity, elsewhere, and restore the sanctity of family, and community solidarity, throughout the globe! Womankind are the ultimate Saviours of all humanity; yet our western sisters must first redeem the egregious, grievous "secularism religion", into a balancing, harmonious living act committed to the God Creator of our American and universal goals of the highest, beneficial options, potentially possible, for our evolvement and beneficent destiny! As God wills! "In Sha Allah"! Where peace, jubilance, equilibrium abounds at last, for all mankind! "Thy will be done, on Earth, as it is in heaven."

CHAPTER 7

"JERUSALEM"

Palestine, configured this century's greatest tragedy: loss of the Palestinian people's homeland for over 2,000 years, and the crux of all subsequent Middle Eastern Arabic Islamic turmoil, civil wars, aggressive, invasive slaughter of innocent uprooted families, in countless millions, now refugees. Palestine, by mandate of the Balfour Declaration Treaty, prior to the Nazi Jewish Holocaust, granted the Jews the homeland nation of Palestine to settle, and occupy, albeit an illegitimate license to dispose of the indigenous Palestinian peoples populace and legitimate lands and properties.

As followed the illegitimate creation of the Jewish state of Israel, by the United Nations decree, and the Zionist world pressures, readily coerced by the Washington Zionist lobby, upon the then, United States' president, Harry Truman. The Arab Palestinians singularly, horrifically paid the price for Hitler's crimes against the Jewish people! Despite having freely given refuge and homeland security to Jews over the countless centuries, as historians are compelled to acknowledge. Jews and Arabs have lived side by side, in peace, eons, recognizing their common Semitic roots. In societal, cultural and historic reality: only Islam practiced non-prejudice, and true brotherhood,

"Umma", for the Jews, as opposed to the notorious biased western European Anglican Christians, and their endless persecutions against the Jewish peoples! When the "Good Ship Hope", laden with Jewish refugees, fleeing persecution, docked in New York's harbor, they were denied embarking by the United States government; only when they anchored in Jaffa harbor, were they welcomed with open humanitarian hands and hearts by the Palestinian Arabs, to find safe refuge! "Our home is your home" - "Ahlan Wa Sahlan"!

An historical event, not documented by pro-Zionist rabid anti-Arab media and propaganda... Islamic peoples, whatever their ethnic or racial lineage, are expressly forewarned by the Holy Prophet of Islam: "All the family of man are one in brotherhood! See ye live and practice Islam's Golden Rule; that ye love for your brother, what ye love for yourself."

Judaism and Jewish inhabitants of Arabic Islamic civilization from the eighth century, over one millennium, reached and realized their zenith and golden era, alongside their illustrious Muslim scientists, literary and philosophical geniuses: rediscovering and creating new distinguished doctors, physicians and philosophers - such as the great Maimonides, Spinoza, and others. Jewish scholarship produced some of the world's most distinguished translations, from the ancient Greeks vast scientific, philosophic sources into Arabic, and Islamic science, astrophysics, medicine, meta-physics, mathematics, biology, oceanography, et cetera, posthumously transmitted to medieval Europe, whose Europe darkling ages foreboding superstitions of fearsome devils, vampires and witches, were over due, ripe and responsive to Islamic scientific enlightenment. Islamic civilization, as reiterated, has evolved the world's most successful egalitarian civilization, never since achieved, with all global man's ethnicity, race, breed and color; where religions of non-Muslims, Jews, and Christians were honored, and their synagogues and churches protected. The Holy Prophet of Islam revealed and founded "the family of man" as one unifying civilization, encompassing all humanity. These precepts must beholden as acts establishing the new world social order of brotherhood, "Umma". Else ye sin, against God and man, and dishonor the holy scripture of Islam. The sacred al Qur'an reveals "God created a ruler, governor, in his stead, man, endowed as "trustee" and guardian of universal nature and all humankind."

The "Arab Spring" did not, all of a sudden, fall out of the blue, nor occur from any western or United States of America concepts of democracy/or so-called equality of justice or liberty, the Arab Spring phenomena: derived from the Islamic legacy and God al Qur'anic revealed religious idiom and precepts; that man is forged and created in God's identifying, Innate Nature, "Al Fitrah"/God; and is endowed with God's very attributes - "Liberty" - "Al Thahir" (the square so named) in Tunis, "Justice"... "Al A'del"... "The Truth"... "Al Haqq"... "The Loving"... "Al

137

Wa'du'de"... "The Sustainer"... "Al Muga't"... "The Infinite Compassionate" ... "Al Rahmin"... "The Beautiful"... "Al Jami'l... "Peace"... "Al Sa'lam"... God's beautiful names and attributes vouchsafed man... of which 99 are known names of the infinite Creator, Conceiver, Beneficent bountiful Originator of man, whilst infinitely more names remain unknown in man's present earth capsule. All man's God identifying names and innate attributes: hence, forward, mankind's uppermost yearning love to aspire to fulfill his own God potential self-realization fulfilling goal as God's Royal Emissary here and now! So, my Western peers, mankind, by virtue of his God creator's vested Divine domain, within: already perceives his or her invested causes for equality, liberty, justice, equanimity, love and peace, sans Western man's concocted secondary "validations"!

"Salem" of Jerusalem – translates as "Peace", and derives from Islam's God Qur'anic name for "Peace", "Al Salam"... Hence, the Muslims' holy name for the city of Jerusalem. Not Jerusalem, the Hebrew word for peace, but the Qur'anic Arabic sacrosanct Name of the God of Peace. The Holy Prophet, enjoined on all Muslims, to greet each other, and all the family of man with "Peace" - "Salam" the language of heaven "Peace be with you" - "Salam ou Alaika" is uttered the Islamic world over to greet and salute both Muslims and non-Muslims.

When the Western Christian crusaders first encountered the Palestinian inhabitants of

Jerusalem, they were greeted with "Peace be with you!" The crusaders were dumbfounded to be greeted by their pre-conceived "enemies"; and shocked to discover a civilization far more advanced than medieval Europe, with civil graces, social justice, scientific enlightenment and highly evolved technically engineered infrastructure. History only tragically records the crusaders' tortuous massacres of thousands of defenseless civilians, Jews and Muslims alike, nor sparing even the Eastern Orthodox Christians.

Palestine was part and parcel of the entire intricate Islamic civilization's lands and peoples, stretching from the Arabian Peninsula eastward through Persia and Asia Minor, and westward through all of North Africa, Morocco, and Spanish Gibraltar. Islamic civilization, as all civilizations, survives, despite military conquests, political suppression, social injustices, but by moral viability and triumph of principles of truth over falsehood, good over evil. Genghis Khan, whose infamous barbarism destroyed Islam's famous Alexandrian libraries, university-mosques, hospitals, and great scientific technical infrastructures, including Islam's astronomical observatories, et cetera, followed by the Ottoman Turks and Colonial powers; yet all horrendous invasions and incursions failed to obliterate Islamic civilization-and its living, viable spiritual veracity and God-centered epicenter of everyday common life. Islam's God-ascribed inheritance remains in Muslims' DNA; and the spiritual live legacy transmitted down generations for over 1500 years, and still inherent, inextinguishable, live in the Muslims' spiritual persona. However, despite all endeavors to "modernize" to secularize, Atheism and irrational consensus of Godlessness is non-existent vis-a-vis our Western mechanistic death of the spiritual identity of man!

Space is too short for the horrific massacres committed by the state of Israel, to the Palestinian peoples, in their "liberation" for Zion Jewry; recorded and documented, not by the pro-Israel lobby and American media, but by the United Nations' world organization sources, (not readily available to the public, except through personal researching the records). Menachem Begin, himself, "the butcher of Dier Yassin", countless other defenseless villagers and indigenous Palestinians; boasts and claims credit for driving out, murdering and assassinating innocents, vulnerable citizenry. The villagers were given five minutes to abandon their homes and lands, or bear the heinous consequences. Whole villages and townspeople slaughtered, the renowned village of Dier Yassin's entire families, women and children buried alive by the "victorious" Israeli armed forces. When the international Red Cross, after repeated requests to investigate the Dier Yassin villagers' tragedy, the victims' bodies had decomposed beyond recognition for their relatives' identification. One Irish playwright, visiting the city of Ramallah, took it upon her conscience to investigate the horrors reached her, by fleeing Palestinian refugees. Ramallah's tens of thousands were forced into the desert. The Irish author, who accompanied the poor miserable people, witnessed herself their desperate plight and thirst driven victims, forced to drink their own urine, to stay alive! She tried to communicate this horrendous ordeal to the western American press, only to fall on deaf ears!

Never mind, ye of heart, humanity and conscience: its over now and the state of Israel, now the world's second Super Power with United States tax-free subsidies, weaponry, and the best B-15 fighter jets, now hold what was Palestine, in the throes of overwhelming military supremacy the Palestinians concurred that a two-state solution would help solve the Palestinian debacle and disaster; reforming and re-instating the original United Nations demarcation lines for the Israeli Jewish state and the original Palestinian inhabitants. Only, even this is rebuffed by Israel, as the majority land is occupied by the Israeli tyrants with no outcry from the international community, least of all the United States of America - who persists to align with pro-Israeli interests, instead of pro-humanity justice! Once, my America and yours, the shining emblem of justice and liberty for all humankind the world over, now supporting and sustaining an occupying, unjust and inhumane Israeli power! When Israel invaded and demolished the peaceful country of Lebanon, no U.S. denunciation; when the still occupier state of Israel invaded Gaza, and committed heinous crimes against the Gaza Arabs, destroying thousands of defenseless innocents, women, children and entire families. The United

141

States of America did nothing to alleviate the untold merciless suffering. Later, however, the international court of "crimes against humanity", was feebly broadcast by the American pro-Israeli media.

How quickly do the Israeli terrorist destroyers of innocent human lives forget they themselves were victims of Nazi heinous crimes? How speedily de-sensitized do Israelis become, in the willful persecution and disastrous destruction of other defenseless human beings, they themselves suffered at the hands of fascist merciless evil? The state of Israel, now perpetuates Hitler's crimes of blatant racism, in a Jewish racist state, the majority rule by the Jewish theocratic Knesset, under the guise of that sacred of human ideologies, democracy! The Palestinian plight of millions could have been somewhat alleviated, had the former 2,000 Palestinian inhabitant owners of their lands, none-the-less have received some humane measure of dignity, equality and humanity: preventing the reactionary enmity and jihadist hostilities, and vengeful murders of both Israelis and Palestinians! Apparently lessons in civil, just humanity, over tyranny, oppression and illegal occupancy, have no human moral stature, in the ethnic, racist visage of the Israelis, who so grievous themselves suffered their homeless Jewish Diaspora!

The gravely misguided Christian fundamentalists and southern Baptists of the American nation misconstrue the biblical reference to the "second coming of Christ" - and "the last days" on Earth, realizing in the creation of the modern state of Israel; hence the staunch, unwavering stance of pro-Israel, against the conspicuous, heinous crimes of the Jewish state against the Palestinian peoples. The Christian scriptures, in no wise infer the injustice of the Holy of holies, Jesus Christ, Messiah, who taught humanitarian love and conscience for all mankind! The gospel prophesied the fulfillment of the "Christ spirit" in all humankind, excluding Judaic exclusive-ness, and world Jewry. Whereas the "Israeli" prophesy, never relegated a physical domain, or portion of "real estate", but the Torah law would fructify in the spiritual kingdom of the living Christ spirit within man's soul domain and heart. Embracing God, the Infinite Creator of Christ Messiah and all mankind. Whereas world Orthodox Jewry is still awaiting their Messiah, still denouncing the Holiest Messiah's greatest world humankind event and triumph! Albeit Islam confirms the blessed Messiah has indeed come revealing truth for all global man, and Al Qur'an reveals "El Messiah" Christ, verily mankind's Saviour sent "as God's very Spirit "Ruah", of the supreme and one God, "self-perpetuating" eternal spirit, "Al Ruah" and Jesus Christ, as "the Compassionate Breath of the Almighty and blessing to all mankind"!

God did not "decree" the modern Jewish State of Israel. God plays no exclusive favorites from the "family of Adam and mankind", but the State of Israel was founded on corrupt power politics and unjust inhumanity to that segment of Palestinian humankind. Zionism was never sanctioned by the world's foremost Judaic leaders; true Torah moral beliefs and tenets were practiced by multiple Jews, including Menuhin, the famous violinist, Ya Ya Hamoudi's father; whose book "The Decadence of Judaism in our Times", was quickly pulled and prohibited from American bookstores. He wrote: "What will we do with the Palestinian peoples, already living there, in their own homeland for over centuries, and numbering millions?" Menuhin's home, in Palo Alto, California, was subsequently bombed inhabitable, by the lawless American "Jewish Defence League".

When the second Khaliph, Ibn Khattab, entered Jerusalem, not a drop of blood was spilt, as he proclaimed to all the inhabitants thereof; "We come, in the Holiest Name of God, the same One God Creator as your God, and all humanity." Ibn Khattab refused his men entry into "the Holy Church of the Sepulcher", honoring the same Christ Messiah of Islam, "El Messiah" venerated by Islamdom. Since that renown time, and holy encounter with "Christian brothers", Muslim families, have historically been entrusted with sacred protection and maintenance of this holy Christian citadel, as posthumously, all Christian churches, as well as Jewish synagogues, protected, by Muslims, in the wake of conquest or immigrating non-Muslims. The famous event and entry into Jerusalem by the Islamic caliph, has finally been published some two decades ago, by

144

the U.S. Time-Life series.

Jerusalem is sacred to Islam as well as Jews and Christians, for the Holy Prophet's disembodied night flight and spiritual ascendency from Jerusalem's celestial sky-dome to Heaven's domain; where El Muhammad encountered all the Abrahamic prophets, Moses. Isaiah heavenwards transported nearest God's exalted Holiest Proximity and whereupon, God revealed to the Prophet of Islam that, he would finalize and fructify God's sanctioned Mosaic-Christ mission for all humankind, as the last Prophet Emissary. The Prophet had perceived a great resplendent light, outshining all others, in Heaven's eternal brilliance and, in answer to the Prophet's curiosity revealed, "This is the light of my blessed Messiah, Jesus!" "O would I have been at Jesus' agonizing death, to have helped Him, from his torturous suffering!" The Ever-Beneficent God replied, "We saved Jesus from all agony – resurrecting him immediately to me and Heaven's angelic host, and leaving a mere likeness of him on the cross! Later, we sent him back to earth, alive, in the flesh, to show our Power of resurrection to his disciples and all mankind."

In commemoration of the Holy Prophet of Islam's ascendency from Jerusalem to Heaven's blessed Stratosphere, Islam built the golden domed "Al Aqsa" mosque, now deemed the holiest site by global Muslim-kind.

How, do we of humanitarian conscience, for

all human beings, without exceptions, solve this Israeli-Palestinian crisis and this century's indisputable tragedy? Any realistic hope lies with Israeli youth, who recognize the Palestinians' equal rights for justice, over their hard liner elders' rigid Jewish orthodoxy. The younger generation of Israelis is war weary, and of military conscript, and yearns whole-heartedly for the future prospect of peace and equanimity with their Palestinian neighbors. They see a bleak future, with their State of Israel, forever hostile, and warmongering, occupying another people's right to self-determination, liberty, justice and peace. The young Israelis know, with unwavering conviction, that there will be no security, no peace - unless justice is achieved for their Palestinian brothers and sisters, equally anxious to build a secure and just future for all! The younger Israelis are more secular and tolerant of diverse ethnic and religious co-habitants, more cosmopolitan in their worldview of humanity, than their ethnic, racial Jewish forebears.

Yet, the greatest single impact on solving the Israeli Palestinian tragic debacle, is the United States of America, who has elected to be Israel's greatest support, morale sustainer ad financial sponsor. The American strictly, pro-Israeli political stance, in no wise addresses Israel's real security and potential prospects for peace, justice and equanimity for Israel's future, in concert with their Palestinian neighbors rights of equal justice and peace-abiding lives. Now and then, a U.S. moral political leader raises his voice to address the moral responsibility of Israel to its Palestinian underlings, like former President Carter, who sees real justice through his moral Christian conscience. A rare instance of daring to speak the truth and underlying cause of' the Israeli- Palestinian tragedy. It seems more politically feasible, however diametrically opposite America's foremost national interests, to invite the Knesset members and spokesmen to align with our United States congress pro-Israeli lobby advocates. We are unequivocally morally corrupt, absurd and contemptible, failing to address both Israel's and our United States highest and best interests, in the name of justice and equality for all, both Israelis and Palestinians. While the Jewish, racist State of Israel perpetuates its illegitimate inhumane occupation, grievous injustices, oppression and crimes against Palestinian humanity: no real peace and justice can be achieved, evermore fomenting and exacerbating the injustices, inequalities, and suppressions inflicted by the status quo present-day hierarchal monarchies, oligarchies, dictators, Egyptian military usurpers and Syria's accursed Assad, warring genocide on his own peoples, and now, causing the 2nd greatest exodus of millions of fleeing Syrian victims, whole families, women, children, grandparent elders, s1nce the Palestinian uprooted millions of refugees. To Europe, risking lives and limbs to Scandinavia, Italy - mostly Germany, whose Angela Merkel's Christian conscience welcomes them... Wherever they can find safe refuge and economic opportunity to care for their families, and educate their children. The Al Qaeda, Boco Haram, Taliban and ISIS terrorists, have only too successfully filled the civil turmoil and massive immigrants desperate exodus vacuum with new heinous crimes, competing with the evil Assad regime we need a new uprising "Arab Spring", implemented by force: since force, is Assad's only methodology to stop his never-ending assault of tyranny and barrel bombing extinction of his own Syrian people. Since the 'Free Syrian Army' and Syrian rebels cause has failed, together with a failed United Nations response; the only salvation is a new revolution, armed by an Arab African -- Tunisian or other integrated Muslim revolt: to triumph and overcome the evils of unrelenting, merciless tyrants. As reiterated, the God-avowed moral veracious tenacity and spiritual resilience for righteousness, human dignity, and man's God imbued attributes for liberty, "Al Thahir", justice, "Al A'del", and humankind's divine innate identifying nature"; Al Fitrah": are all there, inherent, exercised and all pervasive in the Muslim DNA; which can one day realize the establishment and unification of the vast mid-eastern Islamic community. And a new egalitarian replica of Muslim modernity that rids the present dictators, oligarchs and all unjust usurpers of Islamic Unitarian justice and brotherhood - "Umma" of rightful claim and re-instituted justice and equality... For all!

Jerusalem's holy namesake, of "Salaam", or God's name for "Peace", "Al Salam" will, we predict and prophesy, ultimately be the world's shining "City of Peace". Jerusalem is sacred to Jews, Christians and Muslims; the Holy Prophet's

148

al Qur'anic revelation, that, "All mankind is one family," will come to pass, as reality. None shall "own" that Holy Land as private property, but all humanity shall deem Jerusalem theirs; as wilderness and wild life reservations belong to the people, as a sacred trust. Jerusalem, will be open to all... Pilgrims ... Worshiping the one God Creator of all mankind. The United Nations "international city" of Jerusalem, within the decimation, displacement and tragic exodus of the original Palestinian inhabitants thereof, will yet rebirth as Islam's original mandate for "all peoples, creeds, and factions of humanity" to live in peace, harmony and equilibrium.

Israel's Jewish racist ideology will have absolved and converted to the Christ Messiah's universality of all mankind, by way of Islam relegating the worship of Judaism's "One God", over the fallacious "Triune God" of false Christian doctrine. Islam's acceptance and veneration of Jesus as "Christ Messiah": instead of Christianity's worship of the "Divinity of Jesus Christ over the worship of the One True God Creator; will win world Jewry over to Islam. Since Islam encompasses all the Abrahamic scriptures, the Torah and the Christ Messiah's Holy Scriptures, Islam is the Judaic, Christian unifier, and confirmer of all scriptures. Those avid, proselytizing Christian fundamentalists, who proclaim "Christ as God, stand in rank violation of Jesus Christ's own "First Commandment"... That "Ye worship God first, with all thy mind, heart and spirit!" Islam reconfirms the worship of the One

True God, exalted and never equated with human idols, icons or demigods, in Islam's resurrection of Christ's First Commandment of worship of God uniquely and singularly, in Islam's "Shahada... No God but God... God the only Reality"! The Jews of the present State of Israel will have accepted "The kingdom of God is within you," in lieu of a physical locale, or piece of real estate. Ultimately, the forever living God, contained innate within man's God divine endowed legacy, as humankind's God perfecting evolvement fulfills; shall be governed by the Immutable Truth/Reality of God within, over all exterior nationals and man contrived laws! He or she will have attained the Al Qur'anic "perfection of the man species"... "Al Kamal Al Insa'n"!

"Be ye perfect, even as your father in heaven is perfect!" "Thy will be done, on earth, as it is in heaven!"

CHAPTER 8

"THERE IS NO DEATH"

Man, by instinct, rejects his own mortality, regardless of his faith, creed, or religiosity, albeit, agnosticism or ambiguity. Instinct, the base face of intuition that inherent knowledge of Certitude that intuits Immutable Truth, that man knows for certain thrives eternal in his heart, contrary to all argument and superseding overall convention and consensus. Man, intuits his own immortality over mortality, his or her immutable being over his mutable time-spatial earth capsule sequence.

Man's intuiting supra-spiritual sensory being, veritably belies his or her physicality, denies and defies death's finality and destiny! Mankind, by virtue of his own God, intuited, imbued and innate Nature knows He will live forever. For the Orthodox devotee, or religious indoctrinate - who extols his creed, exclusive vis a vie all others: his extraneous letter, sans the spirit and inward living indwelling veracity of God within; falls forever short of God's omniscience, exalted joys and peace of his all Beloving God Creator Cherisher. Man forgets that he is non-self creatable; and reason alone bids he attest an obvious Creator, who makes the likes of you and me, and the inexplicable marvel of universal life-being possible! The man or woman, whose true

151

living faith, already interlinks one, in loving union with God within: who already discovers his or her God endowed innate nature within "Al Fitrah"/ God; stares and dares death in the face, knowing death, but the passage to his higher heaven of evolving being, enlightening truth, peace and love's more singular, exultant bliss! Oh, how we are steeped in our secular fog and oblique impenetrable mist, that masks the finer breath of life and shrouds the Perennial light of Truth, permeating the infinite spiritual reality hidden behind our so finite transitory occupancies and events.

Death in Western culture comes more often, as a sudden shock in the human timely tides of events, too oft, within the secular tenor of our spiritual levity and transitory lives. Among eastern Islamic peoples, death is recognized, not as curtailment, but continued perpetuation of life, from one's earth living capsule of being to life's higher evolving destiny. "Earth is the first growth, and there are eons truth-wards in evolving enlightenment" "Truth", being "God, Immutable Reality ... Al Haqq". The soul of man, "Ruah", indivisibly one with the infinite Soul, "Al Rumi"/God, resurrects immediately at death from its physical prison, soars God-wards to meet with the one God Creator, and "only Reality". ("Al Haqq")

Since Islam recognizes only the eternal spiritual form of man, the body is never viewed at death, in an open casket, never dressed or preserved, but immediately buried, in the earth, the fundamentalist Saudi burial practice, without casket or tomb, is even more eco-friendly for the Earth's perfect recycling operation. Ecologically, it is better for the Earth and man, were human remains, like all lifeless plant and animals, elements and molecular components returned and recycled, sustaining the living earth's balance and equilibrium! Curiously and conspicuously, our western funeral burial customs mimic pagan rituals, more than what should exemplify living Christian faith and optimistic hope for our loved one's eternal bliss. Faith in the hereafter should communicate joy and unswerving commitment to God almighty and our Christ Messiah's truth of paradise to come... Beginning with gracious holy unity of living one with God, here and now:

"The kingdom of God is within you."

Heaven, or "Janah" translates as "Nearness to God", whereas hell or "Jahonum" means "Distance from God". Hell is not a place, but a fire, cleansing and purifying the soul and heart of the man or woman being ... Hell, a state of being, which can be experienced in our earthly tenancy, in the here and now. Hell is not "mortal everlasting" as Christian orthodox doctrine testifies, but as God, the Infinitely Compassionate and ever-merciful God redeemer Wills and Forgives. Hell is depicted in the holy Qur'an, as having rivers of rebirth and renewal of the man soul's purer fibre of being! Methinks, so compassionate and beloving is our Beneficent Creator: that hell, is "God's love in

153

reverse!" God, so multiple times, in Al Qur'an reveals his infinite mercy exceeds his rebuke! I am reminded of a true narrative between a Muslim mystic and God. The Muslim believer communicated to God: "If your worshipers, at large, knew your infinite mercy, as do I; they would never pray so arduously". The answer came... From the One Infinite Beneficent, and forever Merciful, "You keep your secret, and so will I!"

There is that timeless pause and pulse in man, that differentiates between the passing ephemeral and the eternal. That second silent, still breath that brinks on the Immutable ever-living Reality. One split second can expand into measureless perpetuity, and spiritually free and release the imprisoned soul, into infinity of being. "God comes without bell" and Love, un-beholden, adjourns and enwraps and enthralls in pure bliss and peace!

The holier innocence of childhood, not yet entangled with drab, gross conformity, yet lives and signals a supra-sensual spiritual world that imparts timeless dimensions, beyond Earth's tenancy. Still awe-struck by their new marvel of being, these untainted children's pure souls perceive the luminary beauty and wonder of life, from their erstwhile perennial spiritual heaven home. "For such are of the Kingdom of God," depicts Christ of children. Be forewarned, not to meddle and compromise your newborn to our elder's stagnant conventions and bias education.

All babes are geniuses, peering from their spiritual eyes infinitude, whilst yet in their infancy crib. The babe thinks - cogitates, ponders and comprehends more than you can begin to presume; he or she is aghast with over-whelming wonder, in her unique individuality, and as yet limitless unconstrained horizons researching the whys and wherefores of this sudden marvel of life-being! A child's pure heart and natural faith, intuits his God Maker, Creator and divine Father protector. So often tragically compromised or lost with organized religious litany, doctrine or dogma, or worse, by egoistic, atheist parents, secular peers and schooling. "The child is father of the man," penned by Wordsworth, divulges a child's innate wisdom, and nearness to unspoiled nature, God and truth. Such recognition of man's God endowed, innate nature, and being, "Al Fitrah", is the heart of the Islamic revelation, from infancy to man or womanhood. Mankind's God identifying nature and innate God potential for perfecting our man species.

Such a child perceived her own God all-pervasive living reality, at the tender age of three. She easily reasoned that her marvel of life being, derived neither from her own making, nor her parents' creation, but from an ever-living Originator Creator. The author, as surmised, also reasoned that, this life Originator must be necessarily un-creatable and forever alive and eternal; and to my boundless joyous realization, I must undoubtedly share and co-exist forever in this deathless Eternal Being! One can "reason God", as attested in my chapter two, on "the decline of reason in our times "but, dear readers, how blest was I, to reason, in my single solitude, before my speaking age, without props, instruction or influence, other than by God's grace of insightful truth and a pure child's intuiting natural faith! Even then, I in-admissibly rejected my physical mortality for an ever-present spiritual, immortality and immutable reality."

"The pure in heart will see God," reveals El Messiah, Jesus Christ. Albeit, God reveals Himself, as seen in multiple places and faces. In pristine wilderness, wildlife, regal birds on the wing, mountains majesty, or the seas' tempestuous tides, or awe-smitten by the full radiant moon, or celestial stars glorious resplendence; God can be seen in the luminous eye of truth; "Man should learn to detect that gleam of light that flashes in the mind's eye of truth, that imparts genius," notes Ralph Waldo Emerson.

Yet, "the pure in heart", sees God as blinding light's radiance, that blots out and obliterates all physical things and objects of visual reality elevating being to spiritual ecstatic heights that engulfs over and above all else, with God's loving Omnipresence! Such an unforeseen occurrence, experiences pure rapture of joy, as no human agent, or being, man, woman or child is capable of rendering!

The author encountered such an illuminating event - not once, but in several instances, (in the

156

flesh... Now, not "near death", out of body, as others report), but waking, breathing reality, here and now. Mine came, deemed perhaps by the faithful, as "miraculous", assuredly, as God's blessed grace, bestowed on this humblest, insignificant and unassuming one. The author has been most reluctant, to divulge, these events, but through the insistence of those, who believe that, I owe a debt, to my generations, especially in our present-day western egregious and extreme secularism and "proud label" of atheism, which I reverberate as "egolatry", modernity's worship of ego.

The unfolding of these revelations do indeed signify and reconfirm, that, in God's sacred trust and service, I have been vouchsafed a message and mission for my 21st centurions. And I plead your patience and forbearance, in revealing these wholly unsought for, unanticipated events, but by God's holy grace and entrustment.

"Seek and ye shall find... Knock, and the door will be opened," Jesus instructs us, on the undaunted, endless quest for truth and God. Such a never-ending search for God, has ever obsessed and compelled me, from earliest childhood: ever elated, and overwhelmed by this incredible marvel of life and being! The self-propelling search and research followed me through all youth's evolving intellectual, and spiritual aspirations, never answered by schooling or religiosity or conventional status quo.

One early morn, as was my habit to walk in the pure New Hampshire air and surrounding lush green wilderness, brooks, woods, mountains and glistening streams, I had the audacity to ask God, for some intimation of revealing the truth of his, God's Nature, and as His being relates to my own innermost nature of being. I was walking, not towards the rising sun, but westward at that hour, when I was suddenly seized by blinding light, all around me, blocking out the road, on which I walked, and the trees, and all visual objects vanishing. A great radiating light came upon my own head, and I saw my own soul's light merge one in the great blazing canopy of light. I felt overpowering love and over-whelming oneness of rapturous joy, happiness and pure peace, as never encountered on my earthly plane! Praise be unto God, who vouchsafed me some of His Glory, Splendor, beauty and love, I now share with you! All glory and gratitude to God, in his highest beneficence and compassionate grace to one, commonplace as me. I, who have always been grateful for this incredulous gift of life marvel: have been rendered such hallowed grace! I now know the truth of El Messiah's revelation, that "The kingdom of God is within you," available to all, who seek to discover their divine heritage; and came to realize, pursuant to this glorious experience, the confirming truth of the Islamic revelation that "man is forged and created by God within God's own endowed innate Nature." ("Al Fitrah"/God) And we, of mankind possess the God identifying Nature of the All-Infinite Being! Islam also says, "Prayer is converse with God"; thus, my audacious request of God is positively foreknown, already surmised by the Eternal All-knowing, All-infinitely wise, who grants us all knowledge of enlightenment and truth!

Lest I forget, when I returned home my sister greeted me at the door with stark surprise, as she saw my face and eyes all lit aglow. I did not disclose to her what I believed was utmost trust in God's gracious revelation and secret, not to be shared or discussed with anyone person at that momentous time. Years later, when I did, in fact, relate the event with my sister, she wept recalling my light shimmering face. And said she, herself needed this reality of God and faith. My memory flashed back, simultaneously, at that pivotal revelatory experience, at the age of 19, recalling other like experiences of suddenly being overcome with sublime elation, and over-whelmed by a great light's brilliance, at Valley Forge, when left alone, by my family, and finding my way back, the all-consuming light blotting out the trees and paths leading back to the park camp out, and the glorious uplifting joy, love and ethereal peace caught within that luminary moment! I was then a child of 6 or 7 years of age.

I believe that all humanity has access to God's forever-gracious indwelling love. "The kingdom of God is within us." We need only strive and aspire to discover God's Omniscient Presence within us all!

Yet, man still is compelled and driven by his

or her endless quest to know his "Who's Who" identity: his origin, from whence he comes and wherefore he goes. Man ever seeks to know the "Who's Who" Originator, Creator, and Ultimatum of his earthly fate. Mankind ceaselessly seeks to know the Immutable Truth of His Reality seeing He is nothing, Himself, but in that Immutable Reality. And yes, dares visualize and realize that he or she co-shares that eternal deathless reality! That, man lives forever! Hence, man intuits his own immortality, over mortality... His immutability beyond mutability... His infinity over and beyond His finite physicality. "Nothing is physical," saith El Ghazali..."But exists, contingent on its spiritual reality... "No thing or being is physical, but images the light of spiritual immutable reality." ("Al Haqq"/God)

While we may lament the "rose that dies", this beauteous tiny universe, all its own, that blooms so lavishly, images from its spirit-ual reality and beauty's lustrous eternal component, "Ayan" in Arabic. Ayan's spiritual images and immutable components never die. A rose's spiritual consistency lives forever. The famed rose of world poetry, lives forever in its spiritual image, as all pristine beauty on earth's lavishing floral flowering blooms, or as their sovereign Creator wills. For "Beauty", "Al Jamil" is God's name for infinite beauty. True, perfect beauty has its roots in celestial Heaven; Earth is but the physical receptor and evidence of eternal beauty!

Man's eternal quest for the immutable truth

of His immutable Reality, "Al Haqq"/God: is compelled by virtue of his or her own God endowed, innate Nature, - "Al Fitrah"/God. It is, O mankind, your Maker, Originator, Conceiver, Creator, who invites, nay compels you, to seek out your own God-inherent Nature "Al Fitrah" /God - that you know your Immutable Reality. ("Al Haqq"/God) Love and Immutable Truth bid you seek and know your God inspired identifying Nature and Reality. You will never glean the truth of your reality, or the science and wisdom of enlightenment, my western compeers, from your conventions of formal educations and academia associations, with all your PhDs and doctorates. You must enroll in the "University of the Soul". To discover true enlightenment that emanates from man's cognitive soul and spiritual eye that discerns truth from falsehood, the real from the superficial, that images the light of intellect over darkling suppositions morass. Trust your heart that intuits Truth's absolutes... For "the heart is the seat of all wisdom!"

Unlike higher formal education, the "University of the soul" conscripts lifetime continuing education, "graduating" in higher stages of learning, in ascending spiritual intellectual evolving enlightenment, "From the cradle to the grave."

One other wondrous revelatory occurrence blessed this devout seeker of Universal Nature's underlying hidden truth of Immutable Reality - "Al Haqq"/God. Having left New Hampshire's

161

splendor of pristine wilderness, I yearned in my suburban home environment, to encounter again, the spiritual essence of living things - shrubs, trees sanctuary, so frequently found refuge and consoling quiet, in parks, cordoned off public byways. I always carried pen and paper, to record the inspiring bits of insightful truths or poetry, my eyes and ears lent expression to. One autumn day, I paused beside a welcoming lofty sprawling tree, an elm tree, so recalled. I communed with this beautiful living she tree being, and asked, if she would manifest her inner spiritual essence of living nature. Suddenly, her boughs blazed, with burning light, revealing her essence of being! Awe-struck, beyond belief, I myself was enraptured in joyous bliss, peace and eternal gratitude for God's holy revelation. With unequivocal certitude, all living things and being possess this spiritual substance, identity and soul reality! "Nothing is purely physical - but images and emanates from its spiritual Perennial Immutable Reality," reminds El Ghazali. God is the only infinite, all Self-subsisting Reality. ("Al Haqq"/God)

You must divest yourself from all prerequisites of pre-conceived notions or opinions, extricate yourself from conventional expertise consensus, and all populist perspectives to ground zero: above all else, ye must lose thy self-ego to the All-Infinite Mind, "Al Aq'a"/God Originator, Creator, and limitless source of true knowledge and enlightenment. Man is inspired to "hold converse with God", saith the Holy Prophet of Islam and to transcend the mundane, as God's contemplative companion, in illuminative enlightening repartee!

Altogether unforeseen and unexpected: the zenith of my revelatory events, came in a prophetic dream, which I realized, thereafter, held great significance to my human brethren. I saw a vast massive highway before me, the breadth and scope of the earth's diameter-where three human figures, shrouded in shining white, bid me approach them. The first was Jesus, who appeared on my right, in gleaming light, beckoning me to approach him; the second, on the opposite far left, I recognized the Holy Prophet of Islam, who signaled me to join him; the Third Prophet, veiled in white, whom I later realized was Isaiah. All three bade me to join near, and conversed with me, and instructed that I would play a purpose of great magnitude to unite humankind in peace and brotherhood. I fail, my readers, to communicate such a holy prophetic task, albeit, my mission was clearly irreversible ... As one of God's anointed servants to my contemporaries!

I knew in the aftermath of this revelation that I would decisively endeavour to unite the Christian west with the Islamic east: to build a unifying bridge and unification between the human community, and "family of man". I recognized the holy responsibility and sacred trust anointed me; as I know the God of Christ and Islam, are one and the same. And the one unique and singular God of all mankind, enjoins all the family of man realize in these perilous times of needless wars, bloodshed, turmoil, and gravity of injustices, to hearken uppermost to our God endowed innate potential for perfecting peace, loving equanimity, and brotherhood for all humanity! Let us strive to fulfill God's will, in perfecting our God inherent legacy here and now ... To live, thrive and love together in holy peace and brotherhood. "As God wills ... Insha Allah!" Now, while, we still have breath to journey in our so brief earth's tenure!

Our western soulless, Godless mechanistic science depicts us, man, as a "speck" in our solar galaxy, and indeed, in the universe. We are degraded as mere meager nothings... To contend, signify, or vouch for, yet, this insignificant speck, can configure, assess, calculate and ascertain in finite billions, trillions… onwards to infinity. Man, in his integral whole of being, embodies and encompasses the cosmos in his soular infinity! Yea, this mere speck of humanity... Conjugates, contemplates and reflects his infinite God's identity... Vaster than the universe! This 'subject soular seer' sees and comprehends his 'finite object solar galaxies' within his infinite grasp! Albeit man's "soular universe" encompasses infinitude, greater than the finite universe! And who knows the infinite worlds, yet to be traversed in evolving stages of intellectual, spiritual enlightenment and reality, beyond earth's planetary prism!

"Earth is but the first soul's growth, and there are eons of ascending higher evolving stages of truth-wards/God enlightenment!"

164

Nothing dies… Albeit a rose, or any living thing or being, in the universality of nature but images and emanates from its, his or her infinite spiritual and Immutable Reality. ("Al Haqq") There is no death. In the face of pending mortality, man perceives his timeless, space-less, physical-less spiritual reality of immortality. Man or woman, he or she intuits her immutability over mutability. His infinity over time's ephemeral finitude man needs no creed, religiosity, doctrine, cleric or hierarchy, other than God within! His and her latent innate God infused eternity of nature being. ("Al Fitrah"/God) Man knows he or she will live forever.

There is no death!

Love never dies. Love senses an eternal state of being, even here and now. Hence love yields an eternal aural perpetuity of being, while yet on this seeming temporary flux of earthling being. For true lovers, all time inverts timeless, and every moment spans an eternal dimension. Love's sustaining and supportive living liaison and unity of spouses, lovers and companion champions conjures God/Love's Holy innate sanctity of Omnipresence… Love/God is eternal. You will see father, mother, daughter, son, spouse, family kin, lovers, all your lost loved ones in Heaven's link of Love's Eternity. The child of your tragic loss will greet you first, at heaven's gate. Even, your sweet loving animal pets, our bravo loyal canine dogs or majestic feline cats will rouse our sleep. We have "awakened from the dream of life." Albeit every

165

soul, animal or man, evolves higher expanse, greater in God's beneficence of soul, mind spiritual perfecting enlightenment. We shall yet traverse higher enlightening horizons and heights of intellect and heart. New soul's pure loves, and advanced companions, and seers - to share elative new revelations of peerless beauty, buss and iridescent reality!

"Earth is but the first soul's growth and there are infinitude evolvement God-wards, in enlightenment, pure joy's rapture, beauty's radiance and pristine peace"!

"There is no death."

CHAPTER 9

"THE DEMOCRATIZATION OF AMERICAN CAPITALISM"

We have, in earnest intervals, capitalized on mankind's higher evolvement, and spiritual enlightenment of universal truths. How, pray tell, do we even begin to glean and achieve such nobler goals, and inmost inherent God potential for perfection of our species, when "the business of the belly" assails merciless, our planet Earth's vast mass of humanity on a daily basis for dire survival! When global wealth is owned by a few unconscionable gluttons of insatiable greed, the world's economy monopolized and empowered by corporate oligarchic monetary profits that override the desperate human prevailing conditions of destitution and abject poverty: confronting massive worldwide populations. Albeit Africa, India, Asia, of the underdeveloped nations, but encompassing the Western democratic industrial Europe and the United States of America. These imperialist predatory usurpers of Earth's vast resources and wealth plague and subjugate four-fifths of the entire global human race populace.

Economic magnates and their power politics compatriots, who unconscionably afflict desperate economic disparity and horrendous inequality that impact whole generations of families' babes and children into dire squalor and starvation! The global economic catastrophic peril, is headed, and led, no less by our so-called democratic United States republic, and the corporate giants that find tax-free foreign overseas investments, in commercial enterprises, and legal tax-free offshore havens, in the Cayman Islands and other hidden repositories, outside the American domestic economic spectrum.

How has your and my America eclipsed the world's once shining beacon of hope and refuge of millions for freedom of economic equality, welcoming opportunity, security and peace of mind! The once proud American nation of immigrant humanity, escaping economic old world cartel suppression... where has "the American dream" vanished? The epic symbol of social, equitable economic justice and financial security, real property ownership success! The "middle class", once the envy of the world, is now abolished into thin air! The "blue collar" worker, had, in fact, achieved the highest wages for manual labor, anywhere, the world over! Unprecedented opportunity was tangible within reach, sustained by viable financial security, in guaranteed retirement pensions.

The principality of Commerce Corporate ethics actually affected collective conscience, among some of our past titans, in the trade and business. Henry Ford, among other industrial giants, and others, vied and zealously competed with Cadillac, General Motors and other auto companies, for higher wages and benefits, and the conscientious Patron, Ford, proven, in having afforded the common man his own driving vehicle, heretofore only available to the affluent. The vast wealth acquisitions, by America's powerful industrial barons enlisted a "trustee-ship" by those morally conscientious industrialists, that such vast wealth accrued, must be given back to civil society, as "trusted stewards", utilized for the common good, and the "public weal". Henry Ford and Andrew Carnegie, primarily subscribed and applied themselves accordingly, conspicuously, the Scotsman immigrant, Andrew Carnegie, who had acquired the greatest wealth over the others, including Rockefeller, Pews, Morgans, Duponts et cetera. Carnegie gave his entire wealth away, to widows, orphans, and the impoverished - to schools, libraries, and all peace fostering instruments and infrastructures. All just causes "that yield real and permanent good". Carnegie bequeathed his sons, no inheritance whatsoever; only their education, and his surviving widow, sufficient security, only to meet necessitous living costs. Andrew Carnegie, at the culmination of his illustrious career, published his book, "To die wealthy, is to die a sinner!" This uniquely distinct humanitarian, Andrew Carnegie proved a true Christian in spirit and letter, in faith and real life practice. Needless to say, Carnegie's infamous book contaminative of capitalist greed is long since out of print, nowhere to be found. An outright embarrassment, by any comparative paradigm for our other financial wizards, whose "tax-free" foundations far outweigh real Christian benevolence and charity!

How far has American economic enterprise, deviated from yesteryear's Judeo-Christian principles and values of con-scientious practice, with today's American corporate and individual wealth owning the totality of wealth in this American nation: nor longer millionaires, but billionaires comprise the majority of American monetary and property wealth, excluding the off-shore tax havens and compounded interest monies accrued on hidden unaccountable monetary holdings. One percent, or a few hundred individuals or trusts, own $200 trillions dollars of wealth, and the parity average net wealth is $5 billion per billionaire! The Americans comprising the remaining nationals, own less than one percent of America's total wealth, while the 99 percent of America's wealthiest "own" this great nation! How America has once democratic equitable economy digressed and degraded into this contemptible, damnable ratio?

The "middle class" has disintegrated into abject poverty, starvation wages, living from paycheck to paycheck, with both parents' income sustaining bare survival, averaging $35,000, upwards, if lucky to $50,000, yearly at best. The impoverished household income averages $21,000... Food stamps, once deemed shameful, in lieu of preferred work earning capacity, now comprise necessity, in order to supplement the average family's stricken and accursed financial status. Whole families rely on food stamps, free food pantry kitchens and charitable hand-outs, home ownership is now a luxury, millions of families, having lost their prized American home of a lifetime dream, and rentals of shamble homes and apartments scarcely affordable, to sustain a roof, over head and limb, to find refuge from the elements. Homelessness has now become a foul, tragic living nightmare reality for entire families, babes and children's miserable plight, with scores of precious children going hungry! Rank poverty ravishes the land! Where once plentitude flourished for a proud self-reliant independent and solvent American nation!

While America gapes and grapples for a political solution to resolve this, our present abominable economic disparate dire crises: that some real political, moral leadership like a Bernie Sanders or 'Roosevelt reborn', to exact real social, economic justice, once again to American shores, or other means to rectify and transform our present deplorable exploited "free enterprise" capitalist system: there must be inaugurated, a new alternative "democratization" of our present American capitalist system! The author envisages significant social, political, judicial, cultural, economic changes that will translate and elevate our present-day economic business practices, both by government enactment and private enterprises. The latter remedial changes, expressly wrought by the peoples, ourselves - from the bottom grass roots up, for positive achievement and success!

First and foremost: all overseas commercial corporations entities "tax-free" status banned and rescinded, inclusive, as well, their subsidiary tax-free havens, outlawed; and political judicial incentives reconstitute domestic American jobs, quality workmanship and technical skills, in lieu of cheap foreign labor. Jobs that pay decent living wages, whether $8.15 - $20.00 an hour, whatever hourly wage or salary, must be commensurate with the fluctuating G.D.P. standard of living index; and since America, now supports a global economy, our U.S. corporations should apply the same equality of wage earning capacity, as exemplary practice for a new world economic criteria of justice. That would save embattled impoverished workers from gross starvation wages, and alleviate mass poverty and squalor of debasing sub-human living conditions, prevalent the world over!

American domestic and international corporations will pay more than their fair share of taxes, to relieve the present excessive tax burden on the moderate prosperous and poor working class. Concomitant with America's billionaires, now legal tax loopholes and hidden tax havens monetary property and holding stock ventures, declared and accounted for; will pay tax rates commensurate with their income wealth excessive. Since only 200 people own nearly the totality of America's wealth, divided by $173-200 trillions net wealth; these same American magnate family dynasties should be required to pay 50 percent taxes on their ownership each, of at least one-half of America's vast wealth. The proceeds of this fair share taxation, should go directly to mitigate the impoverished, marginalized and disenfranchised millions of Americans and stricken homeless - tens of thousands - in a separate trust foundation, or government auxiliary bank where those unfortunate of American humanity can register and draw whatever meager living necessities will provide basic economic security. These same available funds can be delegated to construct and subsidize affordable housing, to get the homeless off the streets.

American "free enterprise" was never meant or intended to extort and exploit the democratic freedoms of others; moreover on this massive scale of American humanity! As a free license to disuse, abuse and subjugate the masses of democratic American commonality, from their sacred right of individual liberty of economy and opportunity, now stifled and denied by a few egregious greed mongers! The new birthed United States democratic capitalist system was to have created a truly democratic, equitable economy for all the American peoples!

And now, to proceed with the radically new business practices, the author will postulate: in order to secure a true democratizing of our present exploitive American capitalist system, and that will effectuate real positive, transformative change for nationwide economic justice, equality of opportunity, real stability, and promulgate cooperative consolidarity for the vast American nation.

1. First to secure, not only fair and just earning wage capability, for compensating our employees' toil and talents, but grant him or her, an outright interest of ownership in the business, or corporate enterprise. While, there does exist today, corporate businesses and retail outlets, that confer a portion of the net profit proceeds to employees, few, if any apportion all-out capital ownership. Think of the unimaginable incentives and creative incentives that would drive owner participation in such a welcoming venture!

2. Secondly, America's corrupt, exploitive banking system should be entirely overhauled. Wall street's corrupt oligarchic control, dictating political special lobbies, unregulated banks fraudulent fiscal practices that doomed the housing industry and nationwide bankruptcies of millions of proud home owners, must be averted with legislative banking

reforms that serve all the American peoples, not only the powerful few. Our present day banking exclusivity of "credit worthiness" credit or non-credit criteria for good faith extenuating circumstances, should be set and established for new standards creating more equitable systems - to serve all the common people, for both home property ownership and nascent business enterprises. Or yet, private free enterprises sources recreating a "mini-financing" bank, whose resources would invest in the poorest individual and cottage industry skills and talents, to work independently - to secure a living share of the overall economy; like the morally conscientious Bangladeshi, Muhammad Yunus, who has provided interest free loans and economic security for multitudes of impoverished women and their families!

Real fiscal ethical banking reform would dispense with our western Rothschild banking system; that would provide "no interest" bank loans, as envisioned and commenced by the late Arab Islamic Saudi Amir, Faisal, who was tragically assassinated for his robust conscientious reforms, vis a vie contemporary Muslim institutions and civil society; and with an international perspective of all humanity. This prophetic genius and martyr, Faisal Ibn Abdul Aziz el Saud's audacious tenacity of moral reformation were based on and inspired by the holy prophet of Islam's Qur'anic tenets, and reaffirmed in the prophet's final deliverance in Mecca, at Mt. Arafat.

"God has forbidden you, as Muslims, to take usury (interest). Therefore all interest monies over capital investment-shall henceforth cease and annulled (past, present and future). Your capital monies are yours to keep... You will neither afflict nor suffer others any inequity". Strictly incumbent on all Muslim commercial capitalists - as well as wealthy individuals and principals, is to render payment of "Zakat", a fair proportionate "redistribution" contribution of their wealth, to the poor and needy, orphans and widows, as a conscionable obligation, and to ensure the Islamic equitable egalitarian economic, social unification of the Muslim community of "Umma", "brotherhood" is viable and sustainable. In fact, the very word, "charity" does not exist in the Arabic tongue, or any corresponding term. As charity commences in the family homestead, the family father patriarch providing amply, not only for his immediate family, but all, next of kin, sisters, brothers, aunts, uncles, nieces, nephews, cousins even distant relatives, neighbors and strangers, in his outreaching community, are included, if in need. "All the family of mankind", Muslim and non-Muslim, designated by the Holy Prophet, as revealed by the Cherisher Creator, Father of all humanity.

Faisal's new revolutionary, non-interest

bank, did, in fact and reality, become established in Zurich and in other non-publicized innovative Middle Eastern banks, in Bahrain, other gulf banks, and private banks sponsorship. Faisal's Islamic banking system constitutes a "consumer-banking partner" participation, in which the consumer reaps the bank's investment ventures, including investing in the country's natural resources and sound commercial enterprises. Thus, the consumer realizes substantially greater return on his bank monies deposits, and net profits, as "share-holder", not-withstanding the free non-interest loans available to all bank customer consumers. Our United States of America - in adopting this revolutionary concept in banking, would foremost prevent present-day fraudulent practices, of egregious high yields of powerful share-holding monopolistic inequities; and instead, provide the massive commonplace public, a true consumer "fair share" participation, in creating, bolstering and realizing real economic opportunity and independence. This new reformative banking system, necessarily entails monumental political, social, cultural economic change and transformation, from the grass roots up, encompassing the full spectrum of everyday common, ordinary economy, into transformative economic equity extraordinaire! And thus, to uplift the grievous economic plight of our millions of American brethren!

We all know that economic depravity

invariably drives America's violent street crimes, armed robbery assaults, by destitute desperate youths in ever threatening means to secure cash-in retail stores, mini-marts and shopping malls. No public by-way, neighbor-hood or home is safe; from trashed robberies, and burglaries. But, the overall evil culprit that causes and/or accompanies these heinous crimes is America's drug and substance abuse culture, now a national endemic crisis! The latest Oregon campus massacre configures only one of multiple drug-demented psychopaths, who have murdered innocent defenseless victims. Violent crimes only continue to escalate, since the horrific carnage at Virginia Tech, Aurora, Colorado's movie theater, schools and university students' precious lives destroyed, indelible forever in our hearts - the innocent babes murdered in the Newton, Connecticut massacre! Our psychiatric prestigious expertise blame easy access to guns, by deranged minds, as causing America's gun violence.

Drugs and drugs' insidious culture crises cause the root of the gravity of mass homicidal-suicidal murders! Drug and alcohol substance abuse and addiction cause the horrendous continuing perpetuity of our American gun violence and savage crimes, my fellow Americans! Only one of few voices has raised the toxic cause of America's rampant crimes endemic - that of Vermont's governor, Schuler, from whom we have heard the glaring, unequivocal truth of America's

crimes carnage; that drug addiction is pre-
eminently the cause, responsible for today's
national endemic crimes violence!

We return again to the egregious
demoralizing degradation of our American
civilization! Since the pre-historic dawn of
humankind, the family has indubitably necessitated
the implacable unity for mutual beneficial survival
and community con-solidarity. The home and
family are still the basis of moral inheritance,
stamina, and stability that breeds individual and
social conscience, and primarily secures the
homogeneous congeniality and peace of any
community, nation or civilization. The sanctity of
family has demoralized and decimated into horrific
child abuse, sexual assault, and that heinous rape
of incest and all damnable crimes that afflict our
innocence; and these hapless offspring are let loose
on civil society, to commit crime violence, chaos
and murder; and lest we forget, these monsters
devolve, invariably, from a home environment of
drugs and substance abuse, common in such
addicted afflicted families. It matters not, whether
a child is reared by a single parent, or both - the
moral predominance, or lack thereof, constitutes
the deciding factor and reality, in breeding
uplifting, law-abiding citizenry.

The gun violence endemic that terrorizes our
American nation goes integrally hand in hand with
drugs and alcohol addiction, proven to decimate

and derange the minds of would-be murderous psychopaths. The framers of our United States Constitution never intended that every man, woman and child citizenry should arm themselves. The unsettled unpopulated wild west, indeed warranted gun weaponry protection, but once the American teeming viable civilization was secured, the infrastructure of law enforcement, guns were no longer needed, except in real situations that call for gun protection; the Second Amendment of the U. S. Constitution, that purports gun arms weaponry was originally and solely intended for the nascent American nation's protection against the British, and any other foreign invaders. "To carry arms in defence of life and liberty." Preposterous, beyond uncommon sense, that the powerful gun lobby, today, fallaciously and deliberately misinterprets the Second Amendment, and targets obvious stupid and insipid misled advocates, for the gun lobbies own political and financial aggrandizement. And "power politics" propagandized, monopolized exploitive interests over the public "weal", good and safety, and, sadly, too often compromises otherwise political candidates ethics dynamic!

To return to the colossus drug culture, invasive and all pervasive in our contemporaneous American civilization. As aforesaid reverberated: the word alcohol, "alcohol" derives directly from the Arabic language; used mainly as a cleansing antiseptic in medicine and surgery; and drugs – be it hashish, opium poppy, cannabis - were utilized for their medicinal curative properties, and strictly forbidden for recreational use. Alcohol and drugs of any kind are strictly prohibited, as the Islamic scientists discovered alcohol and drugs intake kills billions of brain cells and neurons, and live brain-bodily spinal connecting receptors, disabling and destroying the human immune system. These "mind", bodily vital organs decimating "demons", were therefore serious health hazards and banned principally for medical science's proven mind and bodily disastrous consequences, rather than religious reasons, although Islam's Holy Prophet had warned against alcohol's mind dysfunction and deadly bodily hazards.

Among the international communities of Muslims, who abide by alcoholic and drug prohibitions, an unequivocal anomaly within our global humanity - none of today modernity's rampant diseases of Cancer and other deadly degenerating vascular heart, kidney, liver or other vital organs and glands life threatening diseases, including Alzheimer's disease are prevalent among Muslim-kind. We can, my American brethren, take heed from our Muslim brothers and sisters, to disavow and refrain from alcohol and drugs, and ward off the alcohol, drugs addictions of our youth that paralyze, defunct and crazes otherwise sound and healthy minds! We can be parental commendable examples for our children to follow suit, in pursuit of a "high", subscribe to life's exhilarating, incredulous gift and marvel of being. Seek out your undiscovered latent spiritual mystical jubilance and soul's transcendence, and true "high" rapture. Elate in your soul's infinitive pure ecstasy, in unity with Love/God here and now, in Heaven on Earth, in your own God-endowed divine legacy!

How does the democratization of our American capitalistic system achieve equitable economic reform, when the true wealth of our nation's health: now suffers in dire straits in an unrelenting endemic of cancers and deadly degenerating diseases! When our nation's health is at stake, with over one-third of our population perishing and escalating daily in gruesome numbers! We Americans, withal our past historic three centuries, are now abysmally plagued with pandemic diseases crises; seemingly incurable, by our "medic expertise"; who fail to recognize the root causes of America's tragic disease crises; by the conventional western drugs prescription experimental treatments, with the potential risks of "hit and miss" serious side effects. The ace disconnect is western science's soulless, Godless schism; and medical science's despicable secularist failure to recognize the soul, or spiritual veracity of man - an integral part of the whole medical curative therapy!

3. Universal health care that incorporates a truly holistic soul, mind and bodily prognosis: may be

applied in "Unani", Islamic medical science's original whole soul, mind and bodily medical therapy; that provides our Western disease crises with real cure and preventative disease immunity. "Unani" recognizes (as the ancients) the Soul, "Ruah" as the principal pre-eminent life regenerating, revitalizing force and immunizing "elixir" promoting perfect health, in harmony, alignment and unity with the universal Infinite Soul. ("Al Ruah"/God) As the blessed given regenerating resources necessary for rejuvenation and total health revitalization directly accessible through contemplative prayer and repartee with God, "in elative converse" and ineffable unity with Love/God. For Love and Love's jubilance ultimately heals! The author has covered "Unani" Islamic medical therapy in detailed length in chapter three.

The Islamic organic "Hal" bible foods comprise the optimal health nutritious diet intake - observed and consumed by Islam's international communities Muslim millions, and consumed by the world's healthiest people, the Muslim "Hunzas", with their strictly non-carnivorous, no animal flesh whatsoever. The Hunza diet regime consists of all organic raw live fruit, plant greens, herbs, garlic, and vegetables, whole wheat berries seeds, sesame, pine nuts, lentils, legumes, beans, yoghurt, raw hive honey, and raw nuts, cashews, almonds, and nuts of every variety. This specific diet regime has guaranteed a lifetime disease free longevity from 150-200 years.

Were mankind to truly observe and ingest his sapiens species foods, as designated by universal nature's flawless wisdom, indeed and herein reality, unequivocally proven by the Islamic Hunza peoples, our human species potentiality for perfect health, disease-free longevity, as two centurions old, may yet be attainable and realized. Yet, not without the Muslim Hunzas' spiritual prayers observance of God, five times plus daily: and the Hunza's soul-spiritual vivacities that revitalize and rejuvenate soul, mind and body in prayerful loving unity and peace in God - "Al Islam". The Hunzas' love and omnipresence of God, resonates in loving sanctity of family, joyful optimism, and buoyant inspiring real living faith and community brethren unification! Oh how we, in American civil society, need a spiritual rebirth, renewal and revitalization! Man's spiritual innate God resources, do indeed empower and impact soular, mental and bodily health, over and above all other factors! Contemplative, "telepathic" prayer and "converse with God", do veritably unleash and release spiritual powers and elevating elative transcendent reserves... That jubilate love and peace! For ultimately, God, Infinite compassionate Love/God heals. "Al Rahman Al Rahim!"

Cancer and degenerative diseases are virtually unknown and non-existent among the millions of international Muslim communities. Peradventure, we westerners underestimate the power of prayer, and the ineffable oneness of love that exuberates, renews, purifies and revitalizes! Universal health care and services has been available and free, at no cost, decades now, by the Middle Eastern Arabic Islamic gulf nation's governments, however western cancers and debilitating diseases are rare, unless Muslims have deviated from their organic "Hal" bible foods regime or neglect the spiritual empowering resources of prayer.

Man is not mere animal, but his two-biped being denotes his noble unique stature of dignity of his man species. "The only aristocracy is the nobility of the soul." "Reason" vouchsafed man that elevates mankind over bestial instincts and mind-set limits. "The first thing that God created was reason... That most excellent of man's attributes"! Reveals the Holy Prophet of Islamic Al Hadith. The "aristocracy of man's God nobility of spiritual being", spirituality is man's finer breath of life; man's refinement of his baser sensuality; spirituality is the key that reveals his or her invisible, inextinguishable light of intellect, "Al Noor"; spirituality is the illuminating eye, that pierces the hidden immutable Truth and Perennial Reality behind the universality of Nature. ("Al Haqq"/God) Man's spiritual identity is vital to knowing his "Who's Who" timeless, infinite Originator, Genius Conceiver, Creator: from whom he inherits and derives his God-endowed innate nature, "Al Fitrah", from which/whom, his astounding soul, brain and bodily being, and interlinking divine attributes and capacities for "reason" and rational cognition, from his "Supra Brain", "Al Aq'a", God's supernal infinity of mind. Together with all man's other multifarious "God endowed attributes and qualities conferred on my man creation" "Beauty", "Al Jami'l"/God… "Liberty", "Al Thahir"/God… "Justice", "Al A'del"/God ... Love", "Al Muhabba"… "Mercy", "Al Rahim"/God... "Peace", "Jami'llam", and all man's other ninety-nine known attributes and noblest qualities delegated man, God's elect earthly royal Steward trustee.

Or does man fancy himself, his own "self-created" product? "Why look ye for miracles, O man, when ye are my greatest miracle and marvel of created being!" reveals the Sole unique, ever living, Self-subsisting Maker Creator of all phenomenal marvel of universal life marvel! (Al Qur'an and Al Hadith)

Why does the author reluctantly return to man's higher quest for enlightenment of whom and what he or she is? Because the economic "business of the belly", and the equitable democratizing of our American economy: utterly fails, ultimately to suffice and fulfill man's innate yearning for his God intuited truth of futurity, and blazon insecurity of mutability, versus his or her intuited immutable eternity of being beyond Earth's "bread and breadth" in quest of mankind's higher spiritual evolution and enlightenment are exigent to his personal individual heart and intellectual fulfillment, more perfect evolvement of his species, grossly confronted with our hideous injustices of glaring economic inequality, endemic destitution, poverty and inhumanity; that must cease and desist. And our so marginalised brothers and sisters may recover and realize their core-vested dignity of humanity, amongst new spiritually evolved conscionable generations of American moral tenacity, unification and brotherhood.

Pope Francis visited our American shores, as a moral optimist beacon of hope, in these tumultuous perilous times, for Catholic Christians and non-Christians alike. Addressing the United States Congress and United Nations world body, the Pope agonized over the global prevalence of millions of humanity suffering abject poverty and grievous social injustices, and his powerful appeal to those entrusted with the political, economic, social and cultural echelons of power, whose moral responsibility should take precedence over corrupt and exploitive practices, here in our American nation and the world over. Pope Francis's plea for the "unborn" and citing the evils of abortion, went unheeded by the extreme liberal political unconscionable left, these relentless, unredeemable abortionists attesting their "reproductive rights" overruled any other factors, albeit none would confess to outright "murder" of a human life being, already living, breathing and resuscitating in the mother's womb! These, abortionist women, and their political advocates and congressional supporters: actually define themselves as "Christians"! Just tell the stark unequivocal truth and reality, that you would-be mothers murder your own helpless, defenseless babes. Its "womb infanticide murder", nothing less, by whatever name you conjure or concoct! What if your mothers had aborted you, and your unborn soul and bodily being would never know this miraculous marvel of life being! There are irreversible and universal laws of life and limb, morally accountable for harming or destroying one precious life! And your male advocate friends, will not suffer your insensitive pathetic female accursed unconscionable violations against life's holy sanctity!

While "same-sex" marriages were not openly discussed or disavowed as perverse "sodomy" in Christian truths by His Eminence, the tradition of holy matrimony and the sanctity of family were reaffirmed uppermost. Pope Francis's unwavering insistence that true Christian espousal and faith should perform accordingly, by deeds, indeed presents a real challenge for those who even vehemently profess Christianity, and for political or social compromises, to act totally otherwise! Albeit congressional caucus, the

President of these United States, and his Vice President and the U. S. Supreme Court have enacted "Equal Rights" laws, all must inevitably answer to the Highest Sacred Law! Amen. Pope Francis is indeed a blessed, true servant of Christ, and does praiseworthy honor to the papacy! May God ever bless and reward his beneficent Holiness, with healthful longevity, and God's gracious loving protection!

In the enactment of universal health care, dentistry services should be all-inclusive. The human dental mouth is an integral necessity of overall health, and the wonderful dental teeth, bicuspids, canines, molars, et cetera, so obviously "pre-planned" for their own specific use and function, and is a marvel of engineering - which mankind can only attribute to one unimaginable Technical Genius Conceiver Creator! Our children and youth must ensure that their inimitable natural teeth endowment lasts throughout their lives, so their optimal good health remains sustained. And in-as-much as no substitutes, whatsoever, false teeth, and dentures can compensate for man's highest technical created perfection!

Conspicuously, our elders cannot receive dental care, or sorely needed dentures, under "Medicare", at a time in life, when eminent dental services can relieve and restore much needed tooth replacement. Dentists should mandate the paramount importance of diet, especially foods high in minerals, calcium, vitamin d trace elements, found in raw organic nuts, seeds, legumes and yoghurt. These foods would comprise "preventative dentistry" to ward off plaque deposits cavities and gum disease.

For the environmental health of our earth, and its inhabitants: the agri-industrial farming industry should be banned from further exploitation, abusive corrosive deterioration and destruction of fertile land soil, and wetlands, with contaminant chemicals and pesticides; which deadly poisons, pollute, erode and seep and penetrate into earth's precious pristine water table levels. All the above commercial agri-industrial farming must be outlawed for human consumption safety and health, and for future generations. Therefore strict ordinances should be legislated, for reclaiming the earth's virile, vital ecological balance, and only enlisting organic farming methods, non-harvesting cycles and crop rotation, more suited to the breakup of agri-commercial industries, into local farming and the astute dedication of organic practicing farmers, who will insure the environmental replenishment of the earth's pristine ecology!

The torturous commercial agri-industry's animal farming "manufacture": must also be banned and outlawed. Fowl, and veal, and other animals force-fed, agonizingly stuffed and imprisoned in abusive cages, must cease and desist immediately, not only for cruelty to animals, but harmful for human consumption and hazardous health consequences. Organic laying hens, reared the old fashioned, uncorrupted way, will again reinforce the ecological balance of living natural being, and ensure the health of American humanity! Overall, our marvelous planet earth's present state of lands and seas egregious deadly pollutants and green cases emissions, causing disastrous climate change must be rectified for universal living nature and future humanity. Our seabed's corals, floras, plants and fish are dying. Earth's precious rain forests, with commercial deforestation, and pollutant wind carriers; rain forests of South America and Africa, which provide the whole of humanity with oxygen, vital to living sustainably, are disappearing and dying! We, of the human conscionable common sense intelligence, must impress and impact our politicians and fellow-Americans, of the urgency of climatic upheavals and doomed worldwide catastrophe, unless we act now, with no further threatened havoc that afflicts us all and future generations of humankind!

In a truly democratic capitalist system, education, that now ranks economic privilege for the few, and/or cumbersome hardships of tuition debt for college and university attendance, for our struggling poor, should be cost-free, available to all students. Without access to higher education, individuals and families, can, in no wise secure the economic opportunities and worry-free stability and peace of mind, in the freedom and independent choice of their own profession and career, consummate with one's potential gifts and capabilities. Needless to say, the joy and profounder fulfillment of true achievement!

All education should be free at all colleges, universities, medical, law, science technology, art and all specialized skills and trades, et cetera. All tuition costs, housing accommodations, books, and supplies should be included. How to pay for free universal education, as universal health, America's one percent billionaires, who own 99 percent of our nation's wealth can sponsor and support, by income taxes, in lieu of the shrunken middle class and poor working laborers, who now bear the excessive U.S. tax burdens altogether unjustly!

Once again, Saudi Arabia, Kuwait and the gulf Islamic countries have had sponsored free higher education decades on end - arcane, albeit, to our Western "progressive" mind-set. Whereas, we have reiterated that Islam prizes learning, education and scientific enlightenment first and foremost, proven by Islam's historic cataclysmic sciences that ignited Europe's Renaissance out of the Dark Ages; and which laid the foundation of our own Western technical sciences.

Education and self-enlightenment constitute

Islam's pre-eminent requisite, reveals the Holy
Prophet of Islam, "from the cradle to the grave,"
incumbent on every child, boy or girl.

The now renowned heroine "Malala" who
braved the evil Taliban and survived near death,
could have educated our Western media and
peoples on Islam's quintessential tenet of
education, absolutely incumbent on all Muslims, of
both genders, girls and boys. Islam's true legacy,
not the power dictates of demonic usurpers of
freedom, education, social justice; who defy the
brother and sisterhood -"Umma" of Islamic civil
social unification; who desecrate the holiest name
of Islam, "Al Islam", and God's own name for
"peace"!

The late Amir Faisal would have been proud
of Malala, this amazing Muslim girl. He was
confronted with opposition from opponents, sorely
ignorant of Islam's prerequisites for education, and
historic illustrious legacy of singular scientific
contributions to humanity. Faisal was compelled to
implement by force of army intervention, if
necessary to open schools for girls!

Our Western education, more secular,
composing more surface levity of facts and data,
than profounder latent truths and meaning behind
their limited materialist matrix: should explore and
endeavor to incorporate a more holistic balance of

abstract spiritual metaphysical principles and axioms complement with physical datum; as is integral to eastern Islamic academic, scientific education. Meta-physics should be taught, in tandem with physics, metaphysics is not "philosophy", as perceived by the western mindset, but the "meta" spiritual eye of physics, wholly absent and obscured in present academia and science. The author has reverberated western science's intransigent materialist mechanistic chasm vis a vis the whole spiritual unifying balance and equilibrium of universal living nature. We have traced western science's anti-religious, anti-God debacle to its medieval history of church and theological hierarchal control, over Europe's incipient sciences and discoveries… thence, the disparate irreconcilable sciences and religion hiatus, and western sciences collision course between science and religion, and science's "God schism". We have also underscored the profane spiritual abscess that shrouds the sacred underlying immutable truths, and scientific enlightenment: yet worse, science's de-spiritualizing culture of rank secularist atheism, affecting all facets of life - professional, judicial, social-cultural; and our "medical expertise" drugs proscriptions superficial therapies' abysmal failure to cure the whole spiritual, mental, bodily health of deadly degenerating diseases, and cancers. Patients stripped of their spiritual propensities, and God endowed resources to reclaim and resuscitate precious health of soul, mind and body. Our western, American education curriculum system from earliest primary school should encourage creative writing that excites the power of inmost personal creative discovery of intellectual originality. Parrot senseless, meaningless peer consensus discouraged, and singly personal creativity, given full fledged freedom, to explore each individual's own unique imprints, expression and gifts, endowed by nature. Creativity and language tools: open up the entire human emporium of conscious being, encompassing the spiritual dynamic of literary, poetic and mystical dimensions, that unfurl that inimitable infinite liaison with one's soul "over-soul", as our American prophetic genius, Ralph Waldo Emerson stated. Essayist composing creativity inspires and nurtures profounder creative thinking, expands imaginative and innovating ideas, and ponders philosophical questions. Creativity releases spiritual ethereal propensities that explore his or her own innermost mystical ecology and revelatory light of intellect, and the inspired hearts intuite truths.

Ethics should be taught as an imperative requisite of all formal education - beginning with earliest childhood schooling; for, when not inherited from the home environment, must be incorporated into all schools and education compliance. Only then, can ethical social conscience be fostered, and hopefully adhered to – interacting with civil society.

Comparative world religions should be taught, for their humanitarian moral compass of good will, and peace. And let each student surmise his own God intuite truths of sacred law transcending commonplace mundane secular existence. And so the ardent atheist complainants cease to impose their atheistic nihilism upon our pure, innocent and unbiased youth.

In recreating a truly democratization of the American capitalist system, citing major progressive reforms, beforehand the following four remaining pillars of the new reformed democratic American economy, must be enlisted and enacted.

4. Corporations' overseas tax-free exemptions rescinded, and all offshore tax-free havens illegal. United States banking system completely overhauled, expelling fraudulent fiscal practices, and reformed, to serve the common good of all Americans. Ownership interest participation by corporate and private individual businesses enterprises.

5. Free universal health care, medical and dentistry free services. Free universal education, inclusive of all colleges, universities and technical skills schools.

6. Free electric utilities.

7. Income tax reforms. America's billionaires who own 99 percent of all American wealth revenues will provide all capital funds for the above radical reforms. So the common majority of Americans own an equitable share of America's vast wealth and resources.

We Americans must never underestimate our "peoples' power", our United States of America: uniquely conceived, "of the people, by the people, for the people"! The beauty of true democracy, is that we, the common people can forge our own destiny, and ignite a movement that will create and impact economic equity and equality of opportunity for all Americans poor, destitute, marginalized, and homeless; regardless of ethnicity or race. We are the "human race", with moral imperatives, beholden to our brotherhood of man that can implement all the revolutionary new upscale reforms, in a transformed consolidation of our nationwide communities.

Overall, we need moral leadership from our electives, legislative congress and president, to achieve these significant reforms. Oh, how we Americans sorely need true leadership, another Teddy or Franklin Roosevelt, leaders unabashed and tenacious uncompromising! Who dare confront the political vested powers that subjugate the peon masses; guided by absolute and unwavering faith in the God of our founding nation and the sacred trusts blessed on our American people!

American corporate corrupt business tactics have infiltrated the global economy, as aforesaid alluded, causing the massive peoples of humanity, abject desperate poverty and diseases rampant vulnerability.

Ibn Khaldun, way back in the 9th century, foresaw, as the world's first social scientist, predicted that if man's insatiable greed and lust for economic societal powers were not curbed and cured, mankind's extinction was inevitable. The prophetic scientist compared the analogy of one vessel containing the whole of humanity. If one passenger bore a hole, or fragmented crack, waters would inundate the entire ship, capsizing the vessel, and drowning all the occupants, in fatal death. Ibn Khaldun had foreseen the one day, global economic perils of corrupt commercial greed, as afflicting all humankind. Let us, once again strive to be that hallowed nation, under God, redeeming true brotherhood!

"Like the eagles in the morning sun, a Nation rising from its knees to upset all the histories!" (Anonymous)

Let us lead the global world, for a truly democratic equitable humanity, by the new rebirth of our American civilization! God bless America!

CHAPTER 10

"REVERBERATIONS AND PROPHECIES"

If I have voiced and reverberated, rather redundantly the same underlying message, to my 21st centurion brethren: it is because the overpowering recurring revelatory voice inspires and compels over any feeble utterance of my own. And moreover, being altogether cognizant, that my fellow-man, by virtue of his God endowed inherent nature, can aspire to his highest divine evolving destiny of perfect potentiality, or conversely devolve with admonishing's of humankind's disastrous degradation and dissolution; at this perilous of all foregoing junctures!

These God-inspired summations are not mere opinion; else my trusted sacred task would fall forever short of my communicative message. It seems we must abase our ego and self-esteem to zero, to fructify true, Immutable Reality, intuitively inherent in all of us, of mankind being. That I have peradventure played some miniscule part, in loving service of Love's infinite beneficent, merciful Creator and singular God Cherisher of our man species would somehow justify my own unsolicited gift of existence and life's astonishing marvel!

Western science is only now becoming aware, and in conceding that Islamic sciences – indeed founded and made possible Western technological sciences. While Islam revered and preserved Greek theoretical sciences, it was Islam's transposition of Greco-scientific theories into Islam's genius that radicalized, advanced and uniquely discovered and created new scientific metaphysical principles, and their correlating mathematical disciplines and axioms. The new equations of constants database enabled Islam's geologists to measure with precision accuracy, the circumference of the earth, and their astrophysicists, the apogee of the sun, moon and mercury, in 8th century A.D., before the renaissance of dark medieval Europe had even begun to discover or assimilate Islam's monumental scientific trans-missions to the west. As we have reiterated beforehand: the medieval church's hierarchy of religious edicts, disavowed and denounced scientific discoveries, as secularly irreligious, and it is common historical knowledge, that so many of Europe's early scientists suffered religious persecution and excommunication, and the aftermath glaring chasm between science and religion, that has explicitly caused the seemingly irreconcilable fatal hiatus, ever since.

Whereas we reiterate, Islam never suffered such a divisive religious - science schism: conversely, Islamic revelatory primal precepts held scientists and ordinary mankind to search for all knowledge and scientific enlightenment, beyond Earth's parameters; as all true enlightenment unequivocally led to the timeless, Immutable Truth and Reality. ("Al Haqq"/God) God was not defined by religious litany, but as the infinite Genius Conceiver, Originator of all universal living things and being! Half a century post the Holy Prophet's death, 700 A.D., whilst Europe fancied the Earth flat - Islamic geologists had already confirmed the Earth's circumference, and proven the planet's circular sphere, which later provided the calligraphic maps used by Columbus in discovering the New World and the Americas, only lately acknowledged by historical records.

As Islamic scientists discovered the Earth's oceans and marine life in a new science of oceanography and land geologic rocks, mountains, and valley formations, and Islam's astrophysicists discovered the vast stars and constellations, the universe's colossus of magnificence and endless marvel, their Islamic texts would intercept and pause with wondrous praises and celebration of the unique and one Inimitable Genius Creator and beauty's Peerless Originator!

What a damper on attributing any credence to the Un-creatable God Creator, in Western science, with God absentia from His own universe! All of this astounding marvel of marvels with perfect order and operation between earth, moon, sea tides, sun and the galaxies!! Just "random", accidental... Say Western scientists' rationale? To confess the totality of the universe's flawless governing laws of perfection, is admissible to the obvious, that yea, one of Supernal engineering Genius - Technical Prefectures, hath in living reality executed this phenomenal universe!!

We reverberate, because of Western science's irreconcilable isthmus with religion, and its de-spiritualized doldrums of materialist mechanism: Western science never encompassed Islam's whole science of "Tawhid" - the spiritual and physical substantive entities of universal life, inseparably one compliment to the other. "Tawhid" is the indivisible unity of finite, visible, mutable materiality ("Al Thahir") insoluble with the infinite, invisible spiritual and Immutable Reality ("Al Batin"). As the title of this book explicitly states "Islam's salving science of "Tawhid", would be "salving grace" to Western science's soulless, Godless materialist mechanistic malaise! And would, in turn amplify, expand on and explicate the hidden mysterious dark invisible matter, perplexing our astrophysics-physicists!

We in the West now grapple with pandemic deadly diseases crises: our medical science's drugs and prescriptions therapy abysmally fail to cure Cancers and degenerative diseases. We are in dire need of Islam's medical science of "Unani"; the whole soul, mind and bodily therapy, proven immunity protection for the onslaught of carcinogenic interlopers that disrupt and destroy the human immune system. The Cancer diseases accelerate on a daily basis, destroying one out of every man and every two women's precious lives! America suffers first and foremost from spiritual depravity: the root of our medic's failure to cure the soul and spiritual properties; denied, at the outset, by our western de-spiritualized Godless, secularist sciences. All, in keeping with our American civilization's secularized, demoralized degradation, sapping our moral vitality and spiritual veracity. Our egregious secularist atheist sciences, have contaminated America's religio-spiritual moral fiber, infiltrated and infected our societal, cultural, political mores! Restoration and renewal of our once American veracious spiritual moral character, is nigh impossible, without science's demonstrative transformation, overcoming the science hierarchal spiritual phobias and bias, consensus and persisting fallacies of science versus religion and God-ness. Science, has in fact, emerged as "religion" in its monolithic gospel of Godlessness, reminiscent of yesteryear's church hierarchy, only far worse imposing and contaminating the new "religion of science", in all venues of commonplace everyday civil society, albeit academic, pro-fessional, judicial, economic and to remunerate, medical science, will never take responsibility for America's spiritual paralysis, and near collapse! Ye, the common folk, need a spiritual revolution and higher more sublime transformation, God-wards, re-igniting our nation's founding faith in our Creator Benefactor, Who certified true equality, liberty and the nobler pursuit of opportune fulfillment for all mankind, within brotherhood consolidarity! We, the American nation, stand in dire need for a spiritual regeneration, and the spiritual telepathic power of prayer and unifying repartee with the God Singular Genius Conceiver, Originator of man's flawless human organism - in league with our great historic leaders, whose unswerving faith led us in wisdom and God-wards progression!

Islamic sciences, whose great God-inspired scientific discoveries and enlightenment, bequeathed our own western sciences, have unequivocally proven God vital and exigent to world science's advancement and true enlightenment; God the causal principal factor and intelligential governing laws of Universal Nature, Executor, Engineer and Protector of man's marvel of vital soul, cranium cognitive rational mind, flawless bodily functions and health immune equilibrium!

We must subscribe and re-learn a "whole science" of life, and conspicuously ourselves; our own miraculous living being marvel, and explore the living spiritual underlying realities of our cosmos galaxies magnificence! Islam's science of "Tawhid" reveals the latent innate laws of universal physics indivisible oneness of materiality and spirituality that constitute one immutable reality!

Yet, I will once again endeavour to underscore, in these reverberations, the abstruse incomprehensibility of the Western science and religion schism mind-set, to grasp that the very basis of Western sciences derives from "religion", and its primary sources, of Islam's revelatory precepts and principles of the invisible God creative force factor, all pervasive in the universal living flux and flow of cosmic being entity; energizing the birth of new stellar nebulae and black holes dissolution; God, the inextinguishable Light that emits light to visible sun, Earth, and universal Nature, "Al Nour"/God. God being infinite, (not triune, finite, or flesh) "zero infinity", intangible-from which all quantitative parts, units and numerals formulate. Zero infinity, indivisible, in which have all numerals had quantities ended. Infinity, impenetrable; so even mathematics begins and end with spiritual substantive qualitative properties. Thus, unbeknownst to Western mathematicians, the infinite God factor provocates and produces mathematical sciences, even possible, albeit made accessible to human reason, conferred on man's capacity of cognizance. God's infinite Name translates "Al Kindi", among God's other 99 names, and attributes God is neither finite, flesh nor corporeal human persona. God, the infinite Un-creatable "is not begotten, nor begets, and nothing nor anyone is likened or comparable to God's infinite forever luminous resplendent glory!" revealeth Al Qur'an's second Sura.

Islam is the distinct religious science of the nature of both mutable and immutable Reality, "Al Haqq" - not some whimsical platonic notion, Brahman or Buddhist religious ideology, far reaching mystically hidden per-ception: but here and now, the universal natural world is Supernatural; all mutable ephemeral living things and being are contingent on the Immutable ever living, uniquely Self-subsisting Reality, "Al Haqq"/God. "God/Good is the only reality!" The infinite Immutable God Reality encompasses all-pervasive universal phenomenal being. The cosmos marvel of execution, by supernal intelligential inviolable laws flawless governance, operates by the "will" and irreversible decree of the unfathomable (to us) uniqueness Genius, Beneficent Creator. There are no "random" second, puny guesses, by the likes of us mortals - no coincidence, nor accidental occurrence, as conjectured by Western astro-physicists. "God does not play dice with the universe"! Einstein adamantly attested to   his fellow scientists.

God personally invites his mankind, to observe the cosmic wondrous spectacle: "See ye any incongruity, flaws or blemish in the stars glorious constellations... In the fixed northerly stars affixed to expressly guide mankind's nightly traversing the earth... The sun's rising timing exactitude in earth's orbiting, and its setting... The moon and sea tides' resurgences and calm? Think, O man, reason and reflect," commands man's God infinitely beneficent brain cognizant Bestower of "Reason". "I first created and bestowed reason for my rational man creature!"

All too apparent for those among us, who reason: among myriad ad infinitum: is the Earth's perfect proximity to our solar sun, else the earth would burn up, scorched to oblivion; all this inexplicable marvel of cosmos, "pre-planned" for man and our solar star's incredulous greening photosynthesis for all living sustainable plants, trees, flowering grasses and root food harvesting. "Was there a time, when man was nothing to be remembered?" Aye, but only in the unimaginable mind of the infinitely beneficent, All-beloving, Compassionate. "Al Rahman Al Rahim!" Comes the insipid question, by our astrophysicists: "Is there life out there, communicative, one day to us earthlings"? Be grateful, Oh my fellow-humans, that the overall cosmic pre-conceiver sanctioned for our own protection, millions upon billions of light years distances, to safeguard our Earth home from other galaxies! Otherwise science fiction of nightmares of cosmic and inter-planetary wars, would have become reality! God, "the infinitely All-knowing, All wise", is verily all-cognizant of his untamed, as yet un-evolved human being! "Why look ye for outside miracles and marvels, other than yourselves, O man, when ye are my greatest marvel? Reflect, that I created you, out of nothingness... Imparted my own life spirit... then formed you, from water, earth, and a drop of living sperm... Shaped you in your mother's womb... Gave you seeing, hearing, and consciousness as my knowledgeable creature..." "Al Insa'an" above and over all other beings!" "Ye, O man are the orbital conscience 'seeing eye' central to my universes!" Man conceives, nay sees with his inward God-spiritual, mirroring eye, his affinitive link in the infinitive mind, "Al Aqal"/God, and what an exalting companionable liaison and elative transcending union in man's God infinitive stages of gleaning reflective Light of Truth and Immutable spiritual Reality! "Al Haqq" Here and now, in Earth's present tense, and awesome glimpses of lightening truth, that yet unfurls, enwraps and enrapts in joys of loving oneness; for all truth of reality, ultimately reveals Love/God contained in man's innermost God-infused heart of being. "Earth is the first soul's growth, and there are eons of evolvement ascent God-wards in enlightenment!"

But before we Westerners can even begin to grasp the mind's lightening eye of spiritual gleaning Truth/Reality, we must first balance our infantile and dwarfed preconceptions of universal nature to reiterate Islamic science's inviolable law of "Tawhid" that incorporates both the invisible esoteric, spiritual and visible exoteric material realities, as indivisibly one, as a "whole science of life", albeit redundant, over and again, in appreciation, dear readers for your long patience and endurance in underscoring the eminent significance of the "Tawhid", unity of materiality with spirituality in the universality of all inviolable living thing and being!

Let us restore and revive the "Soul", "Ruah" in our Western clumsy de-spiritualized sciences: contingent on the universal infinite "Soul" of souls" - "Al Ruah" the God, all pervasive eternal timeless factor in our cosmos, in our blind-sided, demented, erroneous, blasphemous science-schism!

We can then commence to cure our deadly western disease cancers pandemic ... Revitalize soul to body-empowering the vital soul, as mankind's greatest "antibiotic" remedial factor! The human soul, "God anointed"… with legions of angelic immunizing antibodies. However or whatever you call these positively credible, intelligential living white cells atoms, that trigger resistance and ward off cancerous interlopers threatening the human body's immune system balance and equilibrium! Who and what causes and impels their marvelous beneficence to man?

Spiritual evolution is vital and inextricable to man's higher perfection. Mankind's perfecting evolvement: necessarily connects and resurrects his spiritual constitution God-wards in reverse magnetic gravity, by elative stages. Man, as such, has attained, reclaimed and realized his divinity of inviolable unity within his or her God inherent innate Nature, "Al Fitrah"/God. Man has realized his immutable reality of being; and henceforth, he relates to his human brethren, through their God omniscient being, as well. There is a supra-sensory telepathic affinity, that now displaces speech; and a light gleams from his brother's glowing soul; and love will smile radiance and joy! Although this perfection of the human species fulfills some 40,000 years from now: it will, inevitably come to pass. Albeit the author retains her unswerving optimism, based on our God endowed blessed nature and true reality of man's God inborn potentiality for perfection of our humankind!

Spiritual evolvement and enlightenment can and does occur even now-amongst our fervent, truly faithful, God inspired, from all world religions' spiritual quintessence, who uplift, adore and celebrate in prayer, and holy unity our earthlings beneficent Creator's magnanimity, love and blessed grace! We, as pure, untainted, unbridled child, can even now experience and joy in "flashbacks" from our eternal home!

Nature requires man's unwavering, unflinching morality, which we live in uncompromising concordance and performance of universal nature's inalienable inviolable laws. The global AIDS endemic, now afflicting millions, is proving more deadly than Europe's bubonic plague, which nearly destroyed the entire human population. AIDS and HIV- infected victims' lives fall fatal in untimely tragedy. Only, and uniquely, the AIDS virus is virtually non-existent among international Muslim-kind. No AIDS or HIV-infected mortals claim victims' precious lives. An anomaly among our human global family, that confounds and confuses with disbelief and consternation among the United Nation's Surgeon General.

The incomprehensible mystery is very simple. Islamic humankind unswervingly upholds the sanctity between man and woman; and uppermost espouses protection family offspring, and the prime dual roles of mother and father complementary parents in rearing sanctioned family unity and unification. Muslims recognize nature and all universal created being, God-endowed spiritual, mental and bodily immune equilibrium, and the irreversible laws of opposite chemical magnetic unity. "Same-sex" affiliations are sexually perverse and counter to universal nature's male and female unity, and destructive to the human immune system's balance and equilibrium, exacting health de-immunizing hazards, which too tragically prove life-threatening and ultimately fatal. The Islamic revelatory laws of inviolable universal human nature are confirmed by Judeo-Christian scripture that denounces any/all sexual perversity of same-sex affinity, as iniquitous depravity, or expressly "sodomy".

Same-sex ideology and practices have sprung from Western secularist populist liberalism, even from those judicial court legislatures, who call themselves "Christians", who totally ignore the dreadful, disastrous health hazards consequences of the neurological-physical chaotic chemical dysfunction and imbalance and disruptive destruction of the human immune system. Where is the World Health Organization, while same-sex practice takes its relentless toll, with the deadliest health hazards confronting the gay segment of humankind? Where are our physician practitioners, who promise false baseless drugs therapies? All for the damning sake of "political correctness"?

We Americans now know the offensive Iraq war was a tragedy that should have been averted, the "intelligence" flawed or not-Saddam Hussein

213

having "weapons of mass destructions", for the unnecessary tragic loss of precious American lives, and the massive Iraqis' upheaval and innocent lives destroyed. Al Qaeda, the Taliban and the other terrorist affiliations, only came into power, with the vacuum caused by the lawless violence and deaths affecting over one million Iraqis. 'Shieh' and "Sunni" sectarian divisiveness, was totally non-existent, only the terrorist enemies exploitive tactics to create chaos and polarized enmities. We have reverberated earlier, that Islam does not, in any wise comprise sects, as is commonplace in western Christendom. The luminary beauty of Islam's egalitarian brotherhood "Umma", unitary unification, was founded by the Holy Prophet of Islam's Al Qur'anic revelation. And in fact, the Prophet commended Muslims to the non-Muslims, diversity of ethnicity and race "as the one family of mankind"! We resonate the historical reality; that Islam was uniquely the world's greatest successful egalitarian non-ethnic, non-racial humanitarian unification, on the face of the Earth, never since equaled, vis-a-vis our Western deplorable, racial injustices! And we can learn from our Eastern Muslim brethren, however our false, fractured American demoralized civilization belies the truth and historical reality against the now predatory heinous injustices of terrorists who deny the Islamic peace-abiding brotherhood, "Umma" of Islam's prime teachings and past historical performance, edification and reality. The terrorist demonical power wheeling horrific crimes, perpetrated and desecrated in the holiest name of "Islam", God's very name for "Peace". "Al Islam!"

Our United States military might better have engaged in Syria, whose evil devil Assad has rained genocide and destroyed Syria, now in rubbles, and causing, out of 24 million Syrian indigenous population, 12 million plus Syrians fleeing Assad's savage atrocities against his own peoples, to emigrate to Europe or any safe haven to live out their battered lives, creating the greatest humanitarian crises, since the Second World War! At least, we could have fully supported and trained the free Syrian army and the resistant rebels, on a more productive scale of feeble, futile half-ass contrivances. Our American Air Force could have bombed Assad's palatial headquarters, with most likely no loss of American lives! We Americans claim to be the world's foremost moral leaders of liberty, equality and justice, in sacred trust of true democratic social, humanitarian justice, yet we failed to abate and mitigate the Syrian fascist tyrant against two million Syrian innocent and defenseless children women and men, whose lives were forfeited by barrel bombs and brutal heinous murderous acts of savagery!

The demons, dictators, and terrorists will ultimately go, as the corrupt, exploitive monarchies, and an illegitimate Egyptian army tyranny, also fall in the wake of another "Arab Spring", not a second, but many more multiple Arab Islamic Springs, revolutionary movements; Muslims inbred God-endowed DNA attributes, summoning yet, once or twice, or thrice again, or however long their social, political, cultural economic transformations endeavor; the peoples themselves, will regenerate their God inherent aspirations for liberty and egalitarian brotherhood and social unification. Where Muslim, Jew, Christian and all non-Muslim minorities are justly represented, in a new egalitarian political order - we in the west term "democracy". Islamic egalitarian "Umma" or "brotherhood" confers on each man, or woman, his or her individual right of equitable justice, not extraneously, as by parliamentary or congressional governance. Each human individual is beholden to the sanction of his own God-innate truth and overall authority within, solidified and confirmed by Islamic communal moral consensus. One occidentals observed that Islam has no religious hierarchy: "Each man is his own priest, or woman, priestess, with obedience to God's sole authority!" The true history of the authentic caliphates of whom the late Amir Faisal attempted to revolutionize and reform his Saudi Arabian compatriots, sadly in vain, I predict will re-adjudicate the kingdom, as will the gulf states pursue the new egalitarian sharing brotherhood unification. The new edifying movements will fire the Islamic soul into vocal political action and resurgence of truly equitable new Muslim communal generations. The thousand years of Islamic kalifates illustrious governance, still vested power in the peoples themselves, according to the rule of the towns, villages and cities populace; where moral veracity governs human conscience and the "collective will of the people", reaffirmed by political jurisdiction, with "authority" to God alone, sans man contrived fallacies; beholden to the commonplace man's personal sanctity with God.

Iran, and the repressive theocratic ayatollahs: will also respond to the Iranian peoples collective will for liberty and social justice. Islam was never a "theocracy" in all its past 1500 years, but as heretofore reverberated, exercised moral veracious liberal egalitarian equality for all the Islamic communities of "Umma" brotherhood. The ancient Persians contributed the quintessential genius to Islamic medicine, mathematics, and monumental scientific discoveries and advancements in scientific enlightenment. Rightly proud of their great achievements! Already Iran has exhibited reformative movements among the majority of Iranians.

As aforementioned, we cannot stress the pivotal quintessential role of religious mores in Islamic everyday, commonplace life. It is near incomprehensible for us westerners to imagine the all pervasiveness of religion in Islamic civil society, in-as-much as western religion, Judeo-Christianity, became an "institution", separate, disparate from ordinary life's secularist spectrum. Conversely Islam infiltrated integral, to all the societal, intellectual, cultural, judicial and scientific sinews of life. We reiterate, that Islamic civil society is the most God-oriented humanity, on the face of the earth! The uplifted transcendent soul of man, God-wards in prayerful communal repartee, repose, and unity, doth breathe a purer ether, and intimate joy in ethereal higher and sublimity of lofty heights. Man's true meaning of "dignity", elevated to new holy horizons! When you walk with God, your Muslim's feet of clay ascend to more sublime horizons!

With the Arabic language, radically transformed by Islam, incorporating God infused words, terms and salutations, as afore reverberated: the daily customary greeting salutations: "Peace be with you" - "A Salam ou Alaikum", addressed both Muslims and non-Muslims; the in vocational remembrance of God, on embarking on any event, travel, or enterprise, vocalizes in "God willing", or "In Sha 'Allah". These are but a few of God's audible prayers and good will, from one brother or sister, or friend, neighbor or stranger. So Islam becomes an inborn legacy, inherent, after generations of hereditary rebirth, and integral to everyday cultural environment. Spirituality, enlivens, uplifts and radiates the human heart with conscience compassion and brotherly unity, equality and peace!

We of the American nation stand in dire, near disastrous need for a spiritual regeneration and unprecedented religious, non-secular reformation: to transform our deplorable spiritual depravity and moral degradation; to uplift our massive nationwide mental anxieties, severe depressions and behavioral bipolar disorders. Once again, we reverberate our secularist science-religious schismatic fatality, which defiles our purer soular angels of our God endowed, higher sanctity of nature. America's lamentable grievous disconnect between God and man. When conscionably resurrected, will revitalize and redeem moral veracity, and eradicate our American nation's abomination of criminal violence, cruel racial injustices, and mayhem murder. For spiritual evolvement in God and brother human, exacts peace, social justice, equanimity and unifying unification for all humanity and earthly co-habitant beings!

We must free ourselves from "selfhood's imprisonment", as Ibn Arabi stated, in 8 A.D.: "Lose thy ego, erroneous and fallacious presumptions and delusions, in search of the truth and 'the absolute' that lies at the heart of soul identity and being infinity." Now opens up the infinitude of immutable truth, and all knowledge and scientia of enlightenment. "I am the infinite all-knowing, the unique and one all wise" revealeth the God Creator source and resources of all knowledge, reason and science imparted to my man being! "Ye must, my human brethren, first shed your self-ego's carnal coil to resuscitate and resurrect your spiritual God vested identity. Wholly in accordance with Islam's primal precept of revelation: that ye must lose all self-ego, "in sweet surrender" to God, thy God Creator Cherisher, in oneness of absolute peace, "Al Islam."

An American spiritual renaissance will never come through organized religion, or sectarian institutions of doctrinal, hierarchal, evangelical or narrow fundamentalist Christianity, that deny the soul of man free and open access to his own divine affinity, sans cleric, doctrine or dogma. Church attendance numbers have declined drastically; as former church devotees have sought a more intimate spiritual consciousness and direct experience, as eastern religions impart. Religion should grow the soul of man, in God-ness conscious-ness. True religious purpose should grow and evolve the soul God-ward in real exhilaration of pure ethereal bliss and peace, not stifled and regress the human soul being, denial of his or her spiritual God evolving journey, through fanaticism and imposing religious exploitive and autocratic control. Once man-simulated hearsay intrudes and mars the pure perfection of God and man's inviolable holy liaison, all is inevitably lost!

Islam, of all the world's religions is without church, temple, or mosque as institution. God is immediately accessible, without icon, saint, Buddhist or Brahman intermediary; there is no intercessor whatsoever between man and God-in direct repartee and ineffable unity with God. Hence, Islam's paramount appeal as the world's fastest growing religion, despite no proselytizing or missionaries conversion, altogether prohibited by Islam. "There is no coercion in religion." One comes solely and freely of his own accord and will, to encounter God directly.

Yea, we of the American segment of humanity, must re-spiritualize, rediscover our soular fire, and once again redeem man's greater good/God within! Only when man realizes his or her indivisible holy link to God's potential limitless being and becoming-will modernity's man begin to dispel rabid madness of crimes against his human brethren; forego drugs for a "religio-spiritual high", instead of the death-dealing disastrous mind's rational destruction, and body's healthful immune system. Supplanting prayerful spiritual resourceful dynamo, versus ego, and restoring, and regenerating the whole soul, mind and bodily

"Unani" of positive, curative therapy; and reconnecting the soul's - "Ruah" – Infinite Soul. And empowering all latent God innate resources. Then, witness men conquer all deadly diseases endemic crises and restoring and resuscitating the wondrous human immune system and optimal healthful, peace-immunizing equilibrium! We already have a century's proven reality in the Muslim Hunza peoples, integrating loving worship of God and universal nature's perfect diet regime, as the world's healthiest peoples; in buoyant boundless health and loving brethren unification, exulting in disease-free longevity of 150-200 years! Oh cherished Beloving, Merciful Creator: liberate and deliver us from our western status quo fallacies, be they of secularist atheist science, cultural or political moral corrupting agendas! Inspire and render man, his or her own true liberty of spiritual, intellectual creativity of more sublime persona of being and may mankind aspire and truly evolve in the enlightenment and truth of his own God innate Nature and Being. ("Al Fitrah") "Man, forged and created in God's own nature and embodiment!" (Sura 30:30) Islam's heart of revelation, for all humankind, intimately realized as man's "royal heritage" and Divine identity. We Westerners can now truly learn the real meaning of "human dignity" attributed by Islam's superlative revelation of man's divine sanctity of unity in God. Man has been "anointed" as Divine emissary for God on Earth, expressly as elect "trustee and steward" of his "Who's Who" sovereign creator in God's own stead. Methinks God trusts us misguided ignoramuses on the sanctified holy grounds of man's inviolable, irreversible nature; ultimately mankind's God innate potentiality for perfect-ion of our species! Else, we must meanwhile rely on God's infinite Beneficent, Compassionate and forever blessed Mercy towards the likes of thee and me, in predictable self-contracted erroneous digressions from God's sacred laws and loving protections!

The Islamic revelatory truth of man's pre-possessed God innate Nature that "salvation", is, by virtue of man's divine link to his or her Nature "Al Fitrah", uniquely endows him or her with God's potentiality for perfection of his species, and coincides with mankind's Christ Messiah's edict: "Be ye perfect even as your father in heaven is perfect"! What a tall order, gone wholly illusive and unheeded by Christendom, conspicuously, because their worship of Christ, precludes Christians aspiring for religio-spiritual perfecting and evolvement of their own souls, and ignoring the Christ's divine message of man's own Godly identity! These worshiping Christians defy and violate Jesus' own "first commandment" - "Thou shall worship the Lord God first, with all thy mind, heart and soul!" Islam obeys Jesus Christ the Messiah's First Commandment, and confirms, "Worship of God first, solely, and singularly", termed "Shahada" in the Arabic. Uniquely, Islam, in holy veneration of Jesus, as the Christ Messiah: deems Jesus, the perfect God Messiah man, "the second perfect Adam" whom all Muslims should emulate on their God ascendency towards perfection.

Al Qur'an, in confirming Jesus Christ, as the only true Messiah, reveals to mankind: "I am the best of saviours... Who sent Jesus, from my own compassionate breath and redeeming blessing for all universal humankind?"

I foresee and prophesy that Islam will indubitably ultimately reconcile western science and religion, with Christendom's reformative religious re-spiritualization, in lieu of church doctrinal litany; Islam will purge western science's narrow constraints and secularist dead-end materialist malaise, and Islam's metaphysical mystical spiritual principles conjoining and complimenting western technical physical stalemates synchronized, as one intrinsic integral "science of the whole" incorporating the finite mutable and infinite Immutable Truth of universal Nature - and encompassing the timeless infinitude of the supernal God immutable recreating and energizing perpetuity of Reality. ("Al Haqq"/ God)

Let us recall for our readers the prophetic words of the eminent Professor Gibb, who espied

Western science's stalemate technical conundrum, and deplored Western irreligious and profane social inequalities and racial disparities: "We must wait upon Islamic society to restore the balance of Western civilization upset by the one-sided nature of European technical progress and to save it from the exacerbated development of European nationalism. No other society (Islamic) has such a record of success in uniting in an equality of status of opportunity, and of endeavor, so many and so various races of mankind" and this triumph of humanitarian unity, at a repressive time, when the Arab Islamic east was occupied by European colonial powers. Despite European colonialization, Islamic moral veracity sanctioned and preserved the unification of brotherhood, "Umma", and Islam's vital cultural mores thrived in everyday commonplace life. Let us recall, as well, the entire Arab Middle Eastern peninsula, all of Arabia, comprised one undivided, non-national terrestrial whole, until England and France divided it into nations and set borders and boundaries. Islam, had for cen-turies thrived as one multifarious plurality of all ethnic and racial diversities, in one contiguous land and sea continuity and unification; Muslims and non-Muslims, as "one family of mankind", revealed by the Holy Prophet of Islam.

Earlier in the book, I quoted Oswald Spengler, who in his "Decline of the West", warned vociferously of the technical materialist de-spiritualized degradation of Western civilization. 'The materialist conception of history, which springs from the same root as Darwinism and like it, kills all that is organic and fateful" Spengler speaks of man's higher providence, which cannot be "cognized", except intuited from within.

Only God/Providence can know mankind's ultimate outcome. Albeit, this author bases her implacable optimism of faith and inexorable certitude on God's infused and innate Nature within man, "Al Fitrah"/God. Neither historic adverse vicissitudes of epoch and generations of man can change or profane: as God's inviolable and immutable sacrosanct omnipresence in man.

The following predictions, or more pronounced tenor of truth, "prophesies" vouchsafed me, as humblest and inspired servant of God: "The infinite All-knowing. All- wise," sole and singular Creator of all man's enlightening science and truth, I address my 21st centurions' generations with certain, positive revelations entrusted me, through God's boundless beneficence and gracious munificence of "grace", or "Il Hymn" in the Arabic.

This book will inevitably ignite, God willing, Western and American enquiry and prolific research and the necessary massive translations of some thousands of Islamic illustrious meta-physicists, and scientists, whose voluminous works have never been translated into Anglo European languages, except for a few rare glimpses, through English, French and German sources. The original sources date back from 7 A.D. through 1400 A.D. Sciences of mathematics, algebra, trigonometry, calculus, and the decimal systems which were, in fact transmitted to Europe's Renaissance, through Jewish scholars translations, laying the mathematical sciences foundations for Western science. However, the original texts contain the abstract, abstruse, spiritual qualitative principles lacking in our mathematics. Islamic medical sciences of optics, surgical methods, plant and laboratory proscriptions therapy, and naturopathic metabolic therapies - Islamic discoveries into all universal nature ecology is too vast for my brief script, encompassing the original sciences of biology, geology, oceanography, zoology, physics, astrophysics, astronomy, metaphysics, mysticism and Islamic religious sources principles and precepts, revelatory and relevant to Islamic scientific discoveries, foundations and monumental advancements, et cetera ad infinitum. We have referred previously to the great Persian doctor's "Canon of Medicine", Ibn Sina, who's "Bible of Medicine" was studied and relied on for seven plus centuries replaced old wives tales and superficial quackery.

We have mentioned before El Burundi's geologic gravitational forces and the earth's orbital circular balance had proven the earth a circular sphere. We again reverberate the signal horticulture agricultural achievements, whose organic farming methods proved incomparably significant to Europe's agri-harvesting: averting yet another European and Anglican famine and rat-infested bionic plague. The piteous persecuted "black cat" feline superstitions were remedied by the Ottoman Turkish ships that transported cats, posthumously proving the majestic cat carnivore, unrivaled success, in mice and rat extermination of infectious deadly diseases and prime deterrent to future infestation, while preserving nature's rodent balance.

These Arabic-Islamist translations, could never serve well than at this ominous timing of western science's fatal materialist abyss of intransigence: and Islamic science's "Tawhid" uniformity and unity of complementary spiritual quintessential property components with materiality. At long last, the glaring absentia in science data, would, not only reconstitute the spiritual dimensions of universal nature-probing more intrinsically, into the underlying truths of scientific enlightenment - but restore, reform and transform a "whole science of life", in lieu of our half-ass materialist dead-end quagmire, going nowhere! Since the whole purpose of science is for greater enlightenment of our astounding cosmos and ourselves, does this not entail plain common sense? By and by, the other monumental significance of cognizing universal nature's spiritual realities, is that we can cure the entire human soul, mind and bodily equilibrium, with the psychic spiritual reserves, and guidelines therapy, that physician's and patients can regenerate to cure our western endemic cancers and deadly diseases crises!

The greatest medical, social and cultural disasters: is America's fatal "drug culture"; proven useless in degenerative diseases experimental treatments superficial therapies, and our American nation's mind decimating, depraving gravity of mental illnesses and behavioral disorders of dysfunction and depression. Medical usage of drugs may be useful, but recreational, is disastrous to the human brain's healthy neurons and the immune system's wondrous protecting resistors fortitude, as we have aforesaid reverberated. If the overall culture of drug addictions calamitous fatalities is signaled by our medical, political leaders, educators and social reformers - America's youth will respond accordingly.

On the escalating AIDS and HIV global disease endemic: only, when the horrifying numbers of our gay, lesbian brothers and sisters reaches staggering numbers, burgeoning into a world-wide crises: will the world health, the United Nations' Surgeon General and diseases center for infectious diseases, finally embark on the medical agenda to alleviate the victims afflicted with the incurable virus. When this deadliest of diseases, like Ebola, that killed one million innocents exacerbate in yet greater numbers: will the medical expertise and humankind, at large, finally attempt to cure the root of the AIDS and HIV pandemic. The root, and unequivocal cause is homosexuality; and finally humanity must confront the inexorable truth. And save their precious lives from certain death!

Nature does not make freaks, but man freaks himself out, with his or her delusive demons. Mankind lacks any freedom, to alter, modify or violate universal nature's inviolable laws. Man is powerless to compromise, in any wise, conversely and exacts the perilous ultimate cost, with his life. The AIDS and HIV diseases will inevitably decease much of global humanity, stricken now in grievous surging numbers. For those obstinate, resigned to paying the dreaded cost with their lives, God have mercy on them. Else, let the others, not as yet initiated in this okay, gay rights culture, be forewarned of the deadliest consequences!

I do happily foresee and prophesy, that mankind's insatiable greed for economic relentless power: will ultimately assuage, with the reformative outcry and morally ascendance of the collective conscience and American and international people's movements for transformation of demoralized capitalist systems. Else, Ibn Khaldun's 9th century warning, which yet echoes through the centuries, of humankind's extinction with unbridled greed, God forbid comes to pass - and exacerbates by unforeseeable calamitous events, before humanitarian morality finally triumphs! I see, poverty, deprivation, hunger and disease ended and extinguished forever, in a global brotherhood and consolidarity of all mankind! Yet, always to implore, "God willing"... In faith of the primal and ultimate Cherisher Creator of such as we, whom God celebrates "as my greatest marvel - man, of all universal life being"! God reveals and communicates absoluteness in confidence of his man marvel being. Should we, not, as well confirm unequivocally, our faith in God's creation?! Albeit it takes we, of humankind, ten millennium, twenty or more, to realize and fulfill our triumphant God-endowed, perfecting beauty of loving brotherly unity, and peace?

The Holy Prophet of Islam underscored mankind's conscientious brotherhood "Umma", egalitarian equality and peace of all ethnic racial unification of man, as "one family". And posthumously, Europe's greatest minds recognized and endorsed Islam's emphatic appeal on peaceful co-habitation with all humankind. "Greet and salute all your human brothers, with "Peace", "Salam", anywhere and everywhere to engender and solidify God's ultimate universal peace! And fulfilling and confirming Christ Jesus El Messiah's... "Thy kingdom come, thy will be done on earth, as it is in heaven!" Islam, being the "Christ Counselor of Truth", confirms and certifies Christ's holy commandments, received from our Father, Creator in Heaven's eternal will for all humanity! Islam's incomparable singular successful egalitarian civilization, over one millennium, and still exemplified by Islam's contemporary international community: my present script, has not the space to include, or yet mention, or quote some of the towering intellects - Goethe, Carlyle, Reverend Richard Frye, Wilfred Cantwell Smith, et cetera; Michael Hames does indeed extol the unifying reality for all humanity. "The central fact of Islam's unity, duality has been put away, no father and son, no division into sacred and secular, or east or west, there is one world, one humanity. It is for this reason that the brotherhood of man is so much stressed in Islam. Racial discrimination cannot exist in such a brotherhood, whether a man's skin is white, yellow, red, brown or black makes not the slightest difference!!

Ronald Ogden recognizes the Islamic

commonality of all mankind, as "dignity" exalted and personified, voicing "Islam exalts the common man, and puts him on the level with the potentate. It derives its inspiration from the spirit. It recognizes the teaching of the Prophet the kinship of all creation. It is essentially democratic in outlook."

We, of conscientious heart, are too oft reminded of the gruesome injustices, so prevalent in our own western American social, cultural, political, economic human conditions. Yet, let us remain optimists, while positive hope does swell in the breast of man's innermost heart of conscience: that ever repels evil for good, inequity for equality, suppression and subjugation for liberty. Man's Good/God inborn, inherent nature and being, has realistically confirmed proven, that the good, Godly forces of his true nature's being, do indeed veritably overcome and redeem his darker ignominious persona! That humankind's elect Divine nature, doth impel him God-wards, in ascendency, in reverse gravity" towards his greater celestial intuiting "Who's Who" infinite all-being.

Humanity's totality of recorded history, does indeed attest victory over evil forces of enslavement, social and economic injustices and tyranny. The present contemporaneous domestic and worldwide terrorism will inevitably be vanquished by the humanitarian conscientious

Good/God empowerment within universal man's faith-abiding struggles. The demon tyrants, Assad, al Qaeda, Taliban, Boco Haram, ISIS, all autocratic dictators, and evil devil predators who harm human innocence, will perish ultimately, hell bound, to answer to the sacred law of eternal justice, love, mercy and peace.

We reverberate and prophesy that man, by virtue of his or her own God innate compelling nature and God potential for perfection of our human species, will fructify peace and brotherhood unification, in a global humanitarian new social world order, (albeit ten, 20, 40, or 400 centuries from now), expedited beforehand by the United Nations mandated "Declaration of Human Rights", with reforms executed, such as the now politically maneuvered and corrupt "security counsel", and allowing the world's majority of nations conglomerate equal voting rights; and to hopefully safeguard global humanity's innocence from horrendous crimes of civil wars and injustices, by recruiting a United Nations army, to implement and exercise force, if necessary, to avert further tragedies of heinous crimes against defenseless humanity. To forever destroy and annihilate all such criminal enemies of the family communities of mankind! In effect, all war on our planet Earth, outlawed! The Middle Eastern Islamic nations, with their resonant greater voice will implement a moral veracity of commonality among our Western rabid secular constituents, however politically, socially motivated.

Let us pray, we of proud American heritage, that our so pervasive fractured de-spiritualized, secularized degradation of our American civilization; of Jefferson, Lincoln, Carnegie and the Roosevelts, retrieves its former God-inspired and founded democratic union and purported "brotherhood" - resurrects in a new re-spiritualizing, morally voracious momentum of rebirth consolidating family, neighbor, and community brethren unity and equilibrium; vanquishing the now marginalised, disenfranchised, destitute and social and racial disparate injustices. True brotherhood, as celebrated in America's intrinsic national character ethics, and resonated in our lyrics anthems and songs, "brotherhood from sea to shining sea!" Finally realized for all our living American masses commonality and future posterity! "God willing", as we perform God's Beneficent Will for all benign mankind brethren!

We happily foresee and prophesy that mankind saves our cherished planet Earth's emergent climate catastrophic chaos, and man made carbon emissions contaminations, will positively avert imminent disaster, and subscribe to alternative energy sources, principally of geothermal, solar and wind power diverting and calming nature's current wrath of weather calamitous tornadoes, floods, earthquakes and violent land and sea erupting catastrophes, and as conscientious man places the harmonious ecological well being of earth and its co-inhabitants before the "dollar" outlawing all earthen mines and fossil fuel's fatal disrupting balance of our planet's orbital harmony, balance and equilibrium. And, in true conjunction, harmonious with universal nature, all poisonous pollutants will cease to endanger all cosmic living things and being, reduced to zero emissions, with America's undaunted resolve, and demonstrable political leadership and audacious moral will!

War, now outlawed, and altogether extinguished, heretofore having ravished, upset, distorted, dysfunctional Earth's orbital balance and integral inter-connecting placid temperance with the solar sun, moon, sea tides and biosphere, stratosphere, in de-oxygenizing live vital ether air currents, will now reinvigorate man and all land and sea amphibians and mammals in perfect biospheric pro-active healthful integral harmony, calm and peaceful equilibrium.

We prophesy that, both Christian and Islamic revelatory scriptures will fulfill in all Universal mankind, will live in peace, love and brotherhood for 1,000 years plus! "The second coming of Christ", is prophesied in Christian scripture and Al Qur'an. While most Christians interpret "Christ's second coming" literally - in the flesh, Islam reveals the second coming of Christ, as the near perfection of all humankind: "the perfect second Adam" consummated and personified in man's perfection, truly "glorify-ing our father in Heaven", or the same and one "God, infinitely Beneficent, Compassionate and Merciful Cherisher-Creator!" as translated by Al Qur'an. "Be ye perfect, even as your father in heaven is perfect," is more heeded by Islam that configures mankind's ultimate purpose and end days towards perfection of man's own God infused potential Nature and Being!

When the prophet of Islam was asked, by one Muslim follower, "How and what do we look for, in Christ's second coming?" The Holy Prophet responded, "You may find him in the spirit of your virtuous brother, perfect in word, deed and acts of brotherly love!"

We, in our dwarfed, inhibiting definitive Western sciences have only reached the "mouth of the cave," in the knowledge and science of enlightenment of our cosmos and ourselves - in our abysmal "relative fumbling's of the lightening absolutes" of spiritual immutable Truth of Reality. ("Al Haqq")

"Who we are," in our incessant quest to

know the intuiting truth of our innermost intrinsic being, ever remains necessarily contingent and veritably exigent on discovering our "Who's Who" Originator, supernal Genius Creator... The Primal Source and Origin of all knowledge, science and enlightenment of being, "Al Il'm". Our inward nature's spiritual God luminosity of eye, whose invisible eternal light penetrates beyond and above our so mundane, carnal and ephemeral finitude. Islam, of all the world's religion, provides both the science and spiritual enlightenment of man's God identifying nature and being, as we have reverberated beforehand. And mankinds illustrious purpose, in realizing his or her highest perfecting evolving being, within our lifetime, in earth's temporal capsule, here and now! And "earth is but the first soul's growth, with eons of ascent truth-wards "Al Haqq"/God in enlightenment, bliss and peace!"

George Bernard Shaw predicted that the era would come, when Islam would challenge, inspire and provide real solutions to the western world's social, racial, cultural, disparities and unresolved complexities, and the West's secularist eroding religious spiritual demoralizing degradation. This modern prophetic insightful genius, espied Islam's unequivocal success as an egolatry of brotherhood, encompassing all diverse ethnicity and races, as our Western world's prime example, to emulate. Shaw recognized the non-institutional status of

Islam, and Islam's appeal to man's own cognizant realization of God's living omnipresence and absoluteness of truths, intuited within man's heart: and the stark amazement of Islam's fastest growing religion over all the world's religions. Shaw attributed, "the already Islamization of European numbers," as significant, in explaining Islam's great appeal to the common man, and liberation from religious hierarchal authority, control and imposed dogma.

The Islamic international nations and communities are the last vestige of humanity, who still retain an integral unequivocal collective sense of the sacred: far removed from our western dual mind-set of life, science and cultural norms, versus religion, and God's all-pervasive unity in every-day life. God is infinitely omnipresent in the here and now, permeating the whole of commonplace life, intellectual, scientific, professional, and judicial and all facets of living. God, love's singular, infinitely beneficent creator is praised and celebrated, from twilight to sunset and in-betwixt. And let us reverberate Ibn Arabi's exultation of God's signal love, and irresistible appeal within the heart of man. "God is the cause of all love. Were it not for love, God would not be worshiped, nay nor adored!"

We have verbalized, reiterated and prophesied Islam's salving alternate solutions and necessary radical reformations of present day western science and medicine: that can positively expand Western science's so dismal, narrow finite limits into Islam's "science of the whole infinitude of universal nature", and Islamic science's "Tawhid" indivisible unity of spiritual esoteric immutable realities with material exoteric mutability. And our exacerbated cancers and degenerative diseases can finally discover real curative results with Islamic medical science's "Unani", the whole soul, mind and bodily immunizing therapies. We, of 21st century civil society: desperately need a spiritual rebirth and realization of our-selves as God endowed, in our potential perfecting evolvement, having no limitations and divinely attributed all possibilities potentialities for perfection of our homo sapiens "man marvel and miracle of being, greater than all my universal marvel of being!"

I still quiver in awe, recalling that beatific vision and revelation, meeting with Jesus Christ, Muhammad, and Isaiah. That I should be so blessed to enjoin in their holiest presence - will forever humble this meager servant of God! Their sacredness of message is clear: that I should unite Christians, Muslims, and non-Muslims under the same and one glorious God Creator Benefactor and Providential Perfector of all humankind! To build a bridge, of transcendent splendor, to uplift, elevate and elate all Universal man being. If my book has served in some humble way to fulfill the Divine will of the forever Beneficent, All-Beloving Merciful Creator, than I am, altogether, indeed humbled and grateful. "To whom much is given, much will be expected," remains a meager understatement, of what must be expected of this blessed recipient of God's magnanimous Love and Holy inspiration! God willing, "Insha Allah", that God guides and yet inspires to fulfill my humblest, loving service to my brothers and sisters!

Oh, Brethren mankind let us right the evils of our world. Let us strive with all our God innate empowering resources to fulfill our highest destiny in loving unity with God, and peaceful brotherly accord with all universal man, and being!

"Thy will be done on earth, as it is in heaven" the Islamic equivalent... "Rahmat Allah"! Or translated, "Thy will be done, to the perfection of all mankind!"

Chapter 11

"ISLAM'S DNA"

MUCH, Yet Much Subsists Writing and Re-writing on Islam, by the So-called Expertise and Muslims themselves. The inbred Muslims cannot see the Forest for the trees. One must be born in our own extremist Secularist Western American Society and Culture, to truly fathom the Impasse of the Muslim DNA.

Islam has never been Institutionalized like our own Judaic-Christian divisions: but Evolved as an Every-day Thriving Way of Life and Culture; amongst the Muslim Global Peoples Themselves,- Bound by Islamic Primal Tenets and Practiceof Religion-Moral Imperatives. Islam's Dynamic of Moral Conduct, Founded by the Holy Prophet, as Brothers and Sisters in Equality, irrespective of Ethnic or Racial Diversity.

"All Humanity are One Family", revealed and mandated by the Holiest Prophet of Islam. We, in the West, can learn from Islam's Brotherhood, "Umma", in lieu of our own Racial Injustices. Despite the Evil Deranged Terrorists, Today, who Defy and Defile the Holy Name of Islam, -which translates "Peace".

Islamic Civilization has endured over 15 Centuries, Sustaining

International Muslims, Living Side by side in Harmony and Peaceful Equilibrium.

The Sanctity of Family, Neighbor, Community, is still upheld uppermost, Muslim and Non-Muslim alike. Marriage between One Man and one Woman is Sanctioned Holy,-Divorce, while permitted, is Rare. No Deadly "Aids" or HIV Infections afflict, Documented by the United Nations, and World Health Organization. "Self-Same Sex" is Deemed Perverse, Counter to Universal Natural Laws of Chemical Unifying Plus and minus.

While Christian Medieval Anglo-Europe was Steeped in darkling Superstitions, vampires, witches and old wives tales, Islamic Sciences, Heralded by Quranic Tenets and Principles,

"To Seek and Discover Universal truths,-though it take Yeti China, beyond,-to the Celestial Starry firmament" Thence the Islamic Scientists Pursued Scientific Enlightenment, Found and Discovered the Basis of our own Modern Technologic Sciences,-only recently acknowledged by Scientists. Islam's Monumental Scientific Enlightenment Ignited and Impacted Europe's Renaissance, now accredited by Modern Sciences.

That Religion could Ignite, Propel and Found Scientific

Knowledge and Enlightenment, forever mystifies Western Scientific perception. Science and Religion remain Irreconcilable to the Western Mindset! The Ancient Greek theories were nullified for the Live, Breathing Universes Scientific Concrete Discoveries. The New Islamic Vast Scientific compendium of Astro-Physics, Astronomy, Moon, Sun and Tides parameters radicalized all the Heretofore Unknown. Islam's Genius of Mathematics Founded both the Qualitative and Quantitative synergy, from the Religious Abstract Muslim axiom, Fused with the

Physical Concrete. Unity collaborating Oneness, replicated God's Oneness. Zero, God's Infinity created the basis of all Islamic Mathematics. "God is infinite, and Greater than its parts". The Decimal System opened up endless Mathematical configures. The Famous Arabic Numerals replaced the useless Roman,-and became the entire Basis of all Mathematical Sciences known Today. Few Westerners even surmise that God's Islamic Infinity has played The Decisive Role in our own Western Mathematical system. Omar Khayyam's Famous Formulas made Possible Mankind's Moon fuller Discovery and Landing.

One Hundred Years after the Holy Prophet's Death, Islamic Geologists had measured the Circumference of the Earth.

The First initial Quranic Surah, "The Clot",
Revealed to the Illiterate Prophet to be...was
"Read, Recite and Praise thy Beneficent Bountiful
Creator Who Created Man from a Clot, and Taught
him by the Pen, what he knoweth not"!

El Muhammed, quivering in Fearful anxiety,
Towering over him, Gabriel, Arch Angel,
Radiating Awe Blinding Light, replied "I am of the
Unlearned and ignorant". "Then Recite in the
Name of thy God Creator,-The Infinite Beneficent,
Merciful and Compassionate"-"Bism Allah Al
Rahman, Al Rahim"! "Ye will Reveal My Truths
to All Mankind". "You, O Muhammed will follow
my Beloved Jesus El Messiah, to Fulfill his
Teachings and My Revealed Commandments"
"You, O Man are My Knowledgeable Creature,
"Al Inssa'n to Rule All Universal Natural beings,
in my Stead".

"You O Man are My Greatest Marvel"! Al Quran
reveals that "The First Thing God Created was
Reason for man". Man is Superfluously Exhorted
to "Think, Reflect, Contemplate and Discover the
multifarious Marvels of Universal Nature". ...."We
have set the Sun in its Course for Man's Daylight,
and the Moon at Night to Guide ye on Land, Sea,
and Fixed Stars for Maritime navigating"......

"We have Devised Time, day, night and Seasons to
Mark the Visible Guideposts, of transient Mutable

flux of Time and Man's Inner Invisible Spiritual Cognizance of Immutable Timelessness in My Infinite Domain"....... "Earth is the First Growth, and there are Eons of Evolving Enlightenment in the Hereafter". Since God is "Immutable Truth",-"Al Haqq",-no Maintained Intercessor or Intermediator, Rabbi, Priest, Cleric Shieh or Human intervention dare enter into that Holiest Repartee of Union between God and Man, Save God. God is Immediately Accessible to Muslims, - through Prayer or Good/Godly Acts. "Dicker", Prayers, Upward and Beyond, Comprise many Daylight and Nighttime Hours. "Remember Praise and Glorify ME, O Mankind, Your Creator, Who Makes your Life Marvel possible". "When My Believer Nears Me, above, beyond his or her Prayers,-I Become his Eyes, through which he sees, her Ears, from which she hears, and his Hands, in My Hand grasp". Man's Identity in God's Divine Omnipresence has More Clarity, than with All Other World Religions. "I have Created Man in my own "Nature",-"Al Fitrah",-Reveals God Revealed in Surah 30:30. ....."In My Own Attributes, Qualities & Beautiful Names". "Love, Beauty, Liberty, Peace. Etcetera Etc...Man is endowed with My Own Attributes that Attest to his own Serene Happiness"

"It Is I, your God Originator Creator Who Forms you in your Mother's Womb, Gives you

Sight, Hearing and Cognizant Mind"...."Formed Ye in Pairs, Man and Woman, to Love, Honor, and Comfort" Apparently, "Adam" was not God's First Creation, followed by Eve....but Both were Created together and Simultaneously. No Snake, Devil or Evil lures Eve,- But God Creates Man and Woman, out of Holiest Love! God Also Implants His Goodness of Conscience in Man. "I have created you with My Own Inborn Goodly Guidance". "God" or Allah", in the Arabic, pronounces "UaLLah" and Translates "ALL INFINITY IN ONE". "Infinity", God, "The ONE" Eternal Living God.

Islam is the Culmination and Fulfillment of the Abrahamic, Moses, Jesus Messiah Commitment and Worship of the One God Creator and Originator of Humankind. Islam Precludes Worship of any/all icons, idols, saints and graven images, Intercessors and Intermediators Other than God, The One Uniqueness Infinity. Jesus Christ is Revered as "El Messiah" by All Muslims, Sent to All Mankind. Indeed, Born of the Virgin Mary, Jesus, Purest in Spirit and in Truth. Jesus is deemed "The Second Perfect Adam" Muslims follow Jesus Christ's own First Commandment "To Worship God" before all others, and accordingly the "Triune god" of most Christians, -as false gods. The Singular Uncreatable God Creator of Man and All Universal Being: is neither

Flesh, nor visceral parts, nor atomical limits of Persona or forms. Al Quran Reveals God is "Forever the Self-Subsisting, Timeless, Space less, and Infinity". Surah 2 Cancels out All and any human devised likeness or associates. "God Does Not Beget, nor Is Begotten, and in

God's Eternal Infinity, Nothing and No one is Comparable". This Key Surah cancels out "the only begotten son" worship of Christendom. Jesus forever glorified "Our Father Who is in Heaven".

Man does not have to "be saved", by a Cruel God. Inflicting Harm and Death on His "Beloved Jesus Messiah". ..."Sent from God's Compassionate Breath and Spirit as A Blessing for all Mankind",-Reveals Al Quran. The Pharisees Jews Committed Jesus to Death, Not The Infinite Compassionate, Merciful God, Father Creator. Even while Jesus suffered on the Cross,-God relieved him instantly, and Raised and Resurrected Jesus to His Glory, Reveals the Holy Quran. Later in Heaven, El Muhammed, the Holy Prophet Saw a Great Light out-shining all others. When Muhammed asked whose Light was this,-God responded,-"Jesus, My Beloved"!

We return to the "Heart of Islam". "God has forged. Created Man in His Own Nature",-"Al Fitnah". Surah 30:30.

We, of the Human Species are "Royals",-

embracing the Common man! Pray tell, what other Religion Elevates Mankind to such Levels of Divine Dignity?

Islam Rejects "Original Sin". God/Love Creates All Living Entity, out of Love. We, of Mankind are "Saved", by virtue of our own God/Good Nature., "Al Fitnah". Heaven and Hell, are not Places, but States of being, experienced in the Here and Now. We can know Hell, suffer Sorrow and Remorse within our Lifespan, or unbounded Joy and Happiness, Intimating Heaven to come! "Sin" has express Billing by Christians, Overridden with Toxic delusions of false conscience. "Sin", Sin and more Sin Driven Self-evoked ultimately end within Incurable paralysis. "God Is The Only Reality" Reveals Islam. "Sin and Evil inevitably perish." God/Good Eternalizes Forever.

"Hell",-"Johanneum", and Means "Distance from God".....and "Heaven", "Jannah", "Nearness to God, States of Mind and Heart. Even "Hell Fire" Serves its God given purpose, Cleansing and Purifying the Wicked....God's Infinite Mercy Abounds in "New Rivers of Re-birth that purify and Redeem".

"LOVE/GOD is the Cause of all love. Were it not for Love,-God would not be worshipped", reveals Ibn Arabi. Islam Invites Man to Hold

249

"Exalted Discourse" and "Silent Converse" with God, in Secret Intimate Repartee. The End and Purpose of All Prayer,-is to Transcend beyond the Mundane, in Exhilarating "Union and Oneness", Sayeth the Holy Prophet, with man's "Who's who" Originator. All true Ardent Prayer Summons "Reverse Magnetic Gravity". Spiritually Soaring Godward.

One Mystic acclaimed his unbounded Love for God, until he Realized God/Love had Preceded his own. Another Muslim Mystic, -Marveling on God's Infinite Beneficent Mercy,-communicated to the Lord God,- that "If your Believers really knew your Boundless, Infinite Mercy,-they would Cease even to Pray.". God Replied,-"You keep your Secret, and so Will I"! "I Was A Treasure Unknown. I Created Man, in order to Be Known", reveals Al Hadith. One Orientalist likened God's Disclosure to this Simile. "The Stars would not know their Glory, Mirrored on the Waters below", reflecting God's Creation of Man, in His Image and Likeness!

"Shariya Law", has been wholly Defiled by the Taliban and Other Terrorists. The Holy Prophet of Islam founded an entire System of Moral Imperatives,-centuries before our own Magna Carter and Bill of Rights. Integrity in all

Human Intercourse, Interaction and Commercial

Enterprises, Required Stringent Ethics Criteria. Food Supplies, Weights and Measurements accorded exact specifics. Property, Assets and Wealth exact Fair Judicial Ownership. Even today, in Modern Times, the Old Shariya Laws are practiced. "Jihadists" conjure false grounds, as well, Vis-a-vie the Savage Barbaric Slaughter of Innocent, Defenseless Civilians. "To Harm One Human being, is to Assault All of Humanity" Reveals the Holy Prophet. "Jihad" entails an Individual Striving Godward to redeem himself or herself in Virtuous Redemptive Acts. What Ludicrous Fallacious Terrorist Crimes, in lieu of Real Jihadist Redemption!

"When God Created Man", revealth Al Quran,- "God Commanded All the Angels to Bow down before Adam"...."All Obeyed except "Sha'atan", or the "Devil",-who boasted he was made of Fire, whereas Adam was made of Clay. Pride Prevailed, Symbol of Ego. In Islam, the Devil has No Real existence, -except as "We Create our Own Devils and darkling unrealities". "The Devil has No Authority over you, O man," reveals God. "I Am The Only Authority and Reality". In Contrast,-the Devil seems to pursue Christians like a "Boogieman", corrupting and exploiting Faith abiding Believers. Islam dispels all Delusions "devil invented" for Unequivocal Absolute

Faith in God, Who is "Truth"-"Al Haqq", now and

Eternal, How We tend to forget the Daily, Nightly "Automatic Involuntary Will" of our Blessed Creator. Albeit, each Living, Waking, Sleeping , Conscious and Unconscious Moment, We of Humankind are "Serviced" by Flawless, Angelic Intelligible Forces and Agents Hidden and Invisible beyond our Microscopic Science lenses: Who Restore, Regenerate and Revitalize Man's Vital Organs and Immune System, Who "Serve" the Likes of you and me mortals!

What Genius of Geniuses Above and Beyond mere human genius, Creates our Vital Organs symposium; our digastric tract's unconscious Stomach synergy,-The Liver's Cleansing Detoxifying powers, Kidneys that excrete exact urea and waste, Intestines miles long, that extricate toxins and Bowels feces wastes,-the endless involuntary Machinations of the Mind embodied neurological system,-with NO Input whatsoever on our conscious part....Astounds Inexplicitly! We are Indeed "Serviced" by Flawless, Perfect "Physicians" Intercessors Loving Tender Care...to boot, "TLC" in man's mundane Vernacular. What Mind-Boggling Marvel of Techno-Engineering Mystifies, All Man's seeming Feats of Inventive contrivances!

"Was there a Time when Man was nothing to be?

Remembered?" Reveals Al Quran. Who and what is Man, that he shares his or her Creator's Divinity and Potential Perfection of Divine Identity of Beauteous qualities and Exalted Attributes of Being,-Man's own God Implanted limitless Creativity of Art, Music, Poetry, Rational perspicacity, Scientific technology! "You, O Man are My Greatest Marvel....Why look for Miracles, Greater than My Man Creation"!

Islamic International Peoples are the Most God- Oriented on the Face of the Earth. We can Learn from their "Umma" Brotherhood of man,-embracing All Diversity of Ethnicity and Races,-Black, Brown, White, Yellow, Etcetera. to Resolve our own Racial Injustices! We can cure our Endemic Cancers with Islam's Medical Science of Whole Soul, Mind and Heart therapy. God,-as the Infinite Youthifying Fount of Spiritual Regenerating Forces that Heal, and Revitalize man's Optimal Health.

The Muslim "Humza's",-the World's Healthiest Peoples have Proven Perfect Mind, embodied Health, Living average longevity to 160-200 Years, Disease Free! To Reiterate,-Islamic Global Peoples Live and Rely on God's Omnipresence, Daily, hourly. There is NO word or Concept for Secular in Arabic. Over 15 centuries of Inbred Religion-Spiritual Liaison,-God is in their DN.

Chapter 12

"SPIRITUAL PRISMS

"Our So-called Expertise go nowhere" OURS, is a Depiritualized Subsistence, Here in our Western Extremist Secularist of Humankind! No Wonder We Fail to Cure our Endemic Cancers, Mental Diseases And the Now Coronavirus that Plagues our United States, Ranking first among all Other Nations. Our So-called Medic Expertise attempt treatments, Mechanistic and Drug therapies, Devoid of Man's Spiritual inner Resources.

We can learn from Mid-Eastern Islamic Sciences of "Tawhid",- The World's First Holistic "Unity" that encorporates Soul, Mind and Body being, as one. Or "Unani", Islamic Medical Science, - Arabic translation, "Immune Union".

Cancers and Mental degenerative Diseases are unknown and Non-existent among the World's Muslim Populations. Since daily Religious activities muster Mind-bodily neuro- charged Immunity against such dreaded diseases. We, inour tragic Cancers and Clinical Depressions, should strive to emulate Islamic Medical Science's Whole Soul, Mind and Emotive Bodily therapy. Although more focus has been drawn to Mental Meditative exercises, inducing Calm: Only Direct Liaison with our "Who's WHO" Mind and Soul Originator Creator can

Ignite and produce Absolute Harmony, Calm and

Sublime Peace, albeit that even Transports to Ecstacy, in "Love/God's Hold in Awe-Wonderful Unity!

However we desist or deny our Spiritual Divine soul in our soul's Soul,-"Al Rah",-we, non-self-created Humans must Attest to Some Magic Miraculous Cause Creator.

"Why look ye, O Man for miracles,-when ye are My Greatest Miracle"! "It is I, Who Shaped ye in your Mother's Womb,- Gave ye Sight, Hearing and Cognizant Mind" Al Quran

"The First thing that God Created was 'Reason' for His mangling",-al Farah",-that Holy Grail. Reason, to Ponder, Explore, Search and Contemplate in his Endless Quest and pursuit of Knowledge and Enlightenment. God's First Quranic Commandment is "To Search the Heavens and beyond Earth for Scientia, Scientific Truths and Enlightenment from the Cradle to the Grave"! Thenceforth, Islamic Sciences Inaugurated True Science, over Medieval Europe's Darkling Vampires, Witches, and Wives' tales fallacies. One Hundred years post the Holy Islamic Prophet's Death,-Muslim Geologists had measured the Circumference of the Earth.

While Greek science was honored,-Islamic

Sciences transposed Greek theories into Real Concrete physical

Data and Axioms. Since the Islamic Revelation Defined "God is Infinite", (not triune), the Zero, "Zipher" hence- forth Laid the Basis for All Mathematics. Numbers, even have "souls", Islamic Mathematics Uniting the Abstract Spiritual Qualitative with Materialist Quantitative. Roman and Greek numerals were replaced by Arabic figures, laying the Basis also, for our Western alphabetical Languages. Complex mathematical Formulas were devised by the Renown Omar Khayyam, which founded the crucial Groundwork that made possible Mankind's exploration of the Moon and Planetary orbits discoveries. The Islamic Geologists and Astronomers proved the Earth Round-when Western Medievalists fancied the Earth Flat. Islamic Sciences exploded Renaissance ignorance into Scientific Truths,- finally acknowledged by Western contemporary Scientists. However Ironic to our Western Mind-set, "Religion versus science" enigma,-in Actuality, over One Millennium of Islamic religio-Sciences Monumental Discoveries have disproven our Western contentious Religious Biases, -our Sciences Derived and Originated from Religious Principles, Tenets and axioms of Islam. Historically Proving the Vital Singular Role that Religion Played, nay Demonstrated in real,

Concrete Scientific Factual Data.

CHAPTER 13

Humankind's Now Coronavirus Plague

Where and how does this most Dreaded Disease afflict Global Mankind? No Cure,-is possible,-no Vaccine yet in the making. As we examine its Origin,-deriving from China,-what astounding peculiarities Cause this Plague?

What is known, is that this Fatal of diseases is carried by an Animal to Human. This Animal must be the Pig, -as the Chinese Peoples are among the World's Populace Who Consume Pork, on a daily basis. The Pig animal, is Nature's worst Carrier of Toxic Bacteria. For this specific Reason, the Hebraic Bible Forbids pig consumption, followed by Islam's prohibition. When 55 Million Peoples Worldwide died of the "Swine Flu",-the orthodox Jews and Muslims peoples were entirely spared. Which glaringly proves the Toxic intake of Pigs, a gruesome Historical Reality.

It is doubtful that any "vaccine" will Work, so long as Pork is consumed. What would Work, your Author believes, is to consume Raw Garlic, since Garlic constitutes Nature's

Greatest, Most Powerful Anti-Bacteria!

The World's Healthiest Peoples are the Muslim Hunzas, who live an average of 150-200 years, Disease Free. Non-Carnivores, they consume No Animal flesh whatsoever. Their Diet consists of Fresh organic Fruits, Veggie's, Seeds, Garlic, Raw Nuts, Whole Wheat grains, yoghurt and Honey. Hunza Women have no "change of life" symptoms, produce offspring after 60 years, and Men Father children over 100 years.

They scoff at our Psychiatric problems,- maintaining their Daily, hourly Prayers, in Liaison with God,-Recharge their Immune system, and Safeguard their Optimal Robust Health. Perchance, we can learn their flawless healthy ways, -in higher than Ego, sublime harmonious Oneness in Creator, since they presumably inhabit our own same Planet!

CHAPTER 14

Spiritual Immunity

The Foregoing has indeed historically proven that Religio-Spirituality acts as the Greatest Safeguard against Mental, Psychic and bodily diseases. Whatever your Prayerful Resources,-We, non-self-created mangling creatures must attest to a Higher Wisdom and Absolute Truth, Governing our puny Mindset, notions and illusive fallacies. Call it God, or our "Who's WHO" Originator/Creator.

Lamentably, our Western American Masses thrive in the World's most Secularist, godless atheist Society,-or among the Extremist Jewish, Christian Fundamentalists. Bigotry Thwarts, stifles our Growth.....and we are here to Grow an Enlightening Soul. "Earth is the First Growth, and there are Eons of Transcendent Evolvement in the Afterlife" Reveals Al Quran.

Ralph Waldo Emerson taught us "To Detect that Glimmer of Light that Flashes on the Intellect" "God comes without Bell"! Emerson revealed cognizance of the "Oversoul",- the "Universal Mind", to which man's Genius is attuned.

Methinks Emerson was privy to Islamic revelations, that connected Man's Brain to the Infinite Universal Mind,-

"Al Aqil". Such a Genius Mind, as Emerson's

must have sought out the World's Foremost Sages. Oddly enough, Emerson, a Renown Pastor of one of Boston's eminent Churches, resigned, in explicit denunciation of Religious Bias and narrow interpretation of God. God is Ever-present within man's Heart of being, Directly communicable,- and ever Accessible. Emerson devoted the remainder of his life,-lecturing on mankind's own Soulful, Mind illuminating Quest for Immutable Truth/God, within.

Emerson's Genius, as America's Foremost Prophet, still goes unbeknown, whilst his protegy Follower, Henry David Thoreau, still reads easier to American literary popularity, albeit profound.

Americans, in direst Straights to personally experience true Spirituality,-persist in discovering multiple Eastern religions, including Buddhism, Zen, Hinduism, etc. etc.

Ironically, despite Islam,-having No Missionaries, or Proselytizing Agents,-Ranks as the World's Fasting growing Religion. "Man's Heart is my Reign and Kingdom" reveals

Islam. "There is No Compulsion in Religion....Man's own Truth of Heart Reveals Me, God".

We must still Overcome our Religious vis a vie

Science Prejudices: Realizing that our Sciences

have originated from Islamic Principles and Scientific Factuality encompassing Monumental Discoveries that Founded our own Western technological Sciences. Islam finally reconciles

Science and Religion. Emerson may have read the Arabic translation on the Inextinguishable Immutable Light that Connects Man's Light of Intellect to God, Who is Light,-

"Al Nour".

"God is the Light of the Heavens and Earth......neither of East or West, but Inextinguishable, Indestructible Immutable Light", reveals Al Quran. In another Sura, God reveals..."I Am the Invisible, Hidden that Fires Fuel to light your lamps and electrify all substance"

Al Burundi, Renown Mystic and Scientist best describes this Perennial Light, as it accords man.

"The Brain Images the Light of Man's Spiritual Eye, Seeing and Mirroring the Spiritual Living Essences of all things and objects and being through that Perennial but Inextinguishable Light "AL Noor" God. Alike, Man's Auditory hearing, with its intricate Canal system Emanates Spiritual Waves that intensify the Epitome of Seeing with being".

Stll, my Fellow-beings, we will never glean the Marvel and Miracle of our Soul, Mind and Body being!

"Why Look ye for Miracles, O Man, when You are My Greatest Miracle", Reveals the Originator Genius Creator of all geniuses!

CHAPTER 15

"EGOLATRY"

Western, Notably American Worship of EGO constitutes Modernity's Religion. "Ego" Masks and Stifles our So-called "Expertise" from truly Probing, fathoming and examining our Latent, profounder Truths. Ego Shrouds the Light of Enlightening Intellect, and delivers hollow, Inane Superficial Consensus amongst our "Expertise". Shallow Levity belies Facts and accumulates limitless factual Data, without any iota of Meaningful Significance!

One of American Psychiatric Geniuses (opposing Freud) likened Mankind's "Subconscious" to God within, Conscient of God or Good. That Man's True Nature emanated from Good or God. When Man violates his God/Good imbued Nature, he descends into an Evil human condition.

Centuries before,-600 AD, Islam revealed "God created man in his own Divine Nature, "Al Fitrah". Sura 30:30. Man is endowed with God's own Attributes, and Qualities of Love, Liberty, Justice, Compassion, Beauty, and All of God's 99 known Names. "I Have Bestowed on Man All of My Beautiful Names and Qualities". No Other of the World's

Religions Distinguishes Man,-as Entrusted Seer and Elected to "Rule and Govern in God's Stead over all Universal Creation". Albeit Truly Evolved Humankind will govern in Goodly Stature, some 40,000 years from now!

Such Unimaginable Gifts and Human properties invest in us, unawares,-taken altogether for sheer grant. Our Amazing Hunan body, firstly. We are Prepossessed by "Involuntary" immoral Movement's,-like Digestion and the Stomach's involuntary Chemical processes, and quick Neutralizing Hydrochloric Acid. The Liver's detoxifying power. Kidneys' Fluids maintenance, Intestines secretions, Lungs' breathing respiratory pulses, and the Heart's Vascular Rhythms Purifying blood,-All these vital Organs in Perfect Synchronism, working together. We are totally unawares and unconscious, every infinitesimal second, that Angelic forces Service the likes of you and me for Optimal Health Immunity.

Our intricate, Complex Brain-Body Neuro-nerves transmitted through Spinal passage, altogether escapes us. The Endless hidden Invisible Spiritual Waves become visible Body/Brain Reality! "TLC" (Tender loving care) unceasingly! This must be a God Creator of Infinite Beneficent Love, Who Renders only Utmost Love beyond which We Humans recognize as love!

Even with all this MAGIC....there are those who fail to recognize Cause, God and Man, Effect. The Undeniable Fact and Reality is that I, and You Exist. I AM. You ARE, therefore God Exists! Only Ego Fallacy denies the Obvious, blinded by Pride and Conceit. Humility, my Peers, Opens Doors of Enlightening Spiritual, Mental Evolvement. We are nothing in ourselves. Lose the Ego,-and Transcend into the Infinite All Wise, "All Knowing". "Earth is Man's First Growth,-and there are Eons Lifeward to Evolve hereafter".

Man, Vouchsafed his Mental Prowess, and Capabilities, can Compute in Trillions into Infinity, within his Universal Space. This is why his or her Soul Encompasses a Universe of Infinite Magnitude. Although Muslim Sufis attest to Out-of-body Ascendency, Islam's Astro-Physicists concur that Man Pulsates a Timeless, Space less Infinitude Within. Humankind Distinguishes between the temporal Ephemeral Now, and the Immutable Serenity of Man's Nature. Again, Islam's Glorious Distinctive Heart of revelation,-that Man is Forged, Imbued in God's Divine Nature, "Al Fitnah".

Sura 30:30.

Islam Reveals "You must Sweet Surrender Self Ego to the Infinite Good/God in Unreserved, Irresistable Love. Lose All self-Ego to the All

Infinite Beloving Creator...Growing, Evolving in Soular, Mind Stature and Being. The Word "Islam" translates,-"Surrender to Peace", "Al Salam", God's Name for Peace.

CHAPTER 16

"LETTER TO MY 20-21st GENERATIONS"

I am Alarmed at our Nation's Widespread Materialist Despiritualized Secular Pulse. Recalling We were Founded under God, our Creator, Endowed with the Righteous Attributes of Liberty, Equality, Peace and the Pursuit of Happiness. Lost is our Faith in God and Fellow Human beings, excepting those few Consent Champions of True Religion Spiritual Democracy.

We are the Sickest of all Global Nations, Our Cancers still afflict untimely Deaths, and Now,-the dreaded Coronavirus Pandemic Plaques our United States Republic Worse than all other Countries! Why,-this Disease Crises, numbering in the thousands, that threatens to cause more Fatalities and Precious lives? Why is our Immunity so Fragile,-vulnerable?

Now hear the Acid Truth,-like it or not. We areConsumers of Alcohol,-which kills Brain Cells in the billions. Neuro-spinal transmitters become paralyzed to energize all Brain, bodily Functions,-including the Vital Heart's vascular Blood rhythmic Flow, the Liver's Detoxicating Power, Kidneys

fluids maintenance, Stomach digestion, Thyroid and all glandular movements, Bowls excretion, etcetera, since All Brain-body Integrate in Unison as One.

The word Alcohol derives from the Arabic "Al-chohal". The Arab-Muslim Physician Scientists who first used alcohol to sterilize their Surgical instruments, Discovered Alcohol destroys billions of vital Brain cells. Alcohol was strictly Banned and prohibited for consumption, thereafter, for Health Reasons, not religious, -hence Islamic Alcoholic Prohibition followed, to Safeguard Human Health. "

We are yet Victims of our Medics, who Deny Man's Spiritual Mind-bodly Empowering Force, for Failed Drugs therapy. We are Deceived by their so-called "Expertise", who Persist in diagnostic physical treatment over Mental Spiritual Propensities. We have aforesaid demonstrated Islam's centuries successful Medical Science "Unani" of Whole Soul, Mind and Bodily therapy, proven Preventative against all Cancers, and Mental Diseases.

Mankind may yet evolve to Perfection of our Species,- albeit 20 Millenniums from now. Witnessing his or her Individual insatiable Greed and Global Corporate RuthlessEconomic Exploitation.

America once led Priority in Ethical Business Values. Founded "Under God", -Propagated by

Henry Ford and Andrew Carnegie,- the True Christian Criteria of Economic Justice was practiced.

There was a time, when Corporate Monopolies were outlawed,-until the U. S. A. abandoned Judaic-Christian Moral Principles of Equity and Economic Justice, "Free Enterprise" Opportunity accessible to the Common, Middle Class Man., - now oppressed by Corporate Greed Control and Stranglehold. America's once Massive Middle Class, now Extinct,-and National Economic Deprivation Stares in the Face of American Majority,-whilst America's vast Wealth is Owned, manipulated by Billionaires. Welfare, once scoffed by hard-working Americans,-now Depend on Food Stamps and hand-outs. America's Total Wealth is now owned by 99 percent of the Privileged, while 1 percent is retained by Commonplace American families, desperately Striving to secure Life's barest needs for Survival!

Bernie Sanders has dedicated and committed his Life's Political endeavors to Stamp out Corporate Insatiable Greed of these Vultures,-in his would be "Political Revolution"...... To achieve "Free Universal Health Services" (as do All others of the Industrial Democratic Nations) Economic Justice, Curbing American Corporate monopolies' Avarice. Senator Sanders has succeed in Winning over millions of Millennial's to his Vision and

Humanitarian Leadership. Alas, his withdrawing his Presidential candidacy, leaving Biden, with his past performance of inept, status quo Baggage,- will assuredly provide Donald Trump, his second Term of Evil, Corrupt Presidency! While Biden masks under Obama's Virtuous Domain,-No iota of meritorious Comparison exists! Millions of Democrats will never Vote for the Biden Pretender!

"Be ye Perfect, even as your Father is perfect" still goes unheeded by Christian Majority. What a Command, Imparting Mankind's rightful Divine Identity, as Children of God, our Heavenly Father. Jesus Christ, Messiah's First Commandment was to "To Worship God First, with all thy Soul, Mind and Heart". Islam, who Venerates Christ, as the True Messiah,-Practices Christ's First Commandment,- Denouncing all Icons, images, idols and statutes for the One Divine God Creator Who "Reigns in the Heart of man". Verilt, "The Kingdom of God is within you".

The "Brotherhood of Man" has historically gone unheeded with our Western World's grave Racial Injustices. Christ's Vision of Brotherhood, became Reality in Islam's Egalitarian Brotherhood of Mankind, "Uma", in True Ethnic, Racial Equality of All Diverse Plurality of Humanity. Despite the Arab Mid-Eastern Evil Dictators, Oppressors, Terrorists,- the Peoples themselves

Flourish in Islamic Brother- Sisterhood in Moral tenacity and Live Cultural Reality! "All Humanity are One Family"-revealed the Holy Prophet of Islam. Quite unknown by We Westerners, is that Islam Reveals, Muhammed to be the "Comforter" and "Counselor of Truth", Jesus Christ prophesized.

Islam's "Father of Sociology",-Ibn Calhoun analogized All Mankind's Survival and Destiny, likened, akin to All Humanity being Afloat together in One Boat,-if and when All beings Ensure Peace, by Loving Cooperation and Unity as One Species,-or Fall from Good/God's Grace to Disaster and Extinction.

Chapter 17

ISLAM ---

From the Point of View of an American Born.

My Pride lies in our Founding American Fathers, Who Created A Singular Constitution,-based on Mankind's Equality, Freedom, Justice and the Pursuit of Happiness.

To Entrust all Future Social Events, Premised on our Creator God's Beneficent Will for all Humankind, "Under God", even published on our Dollar, Five, Ten and Twenty Dollar bills,

The Vital Union of Indivisible Unity was upheld once again, by Abraham Lincoln, in a bitter disastrous Civil War, Reaffirming All Men and Women's Inalienable Right to Liberty and Judicial Equality. Subsequently,-All Civil Racial Rights have been fought and legislated into Law. Despite America's struggles for Racial Justice,-our American Black Community continues to Suffer Prejudice and injustices!

We can Learn from Islam: Who Freed Everyman from Ethnic, Social Injustices,-Establishing The World's First Successful Egalitarian Society, in a Vast, Plurality of Ethnic and Racial True Equality. nearly 18 Centuries ago, Called "Umma", A Real Brother-Sisterhood of

Individuals and Communities. Despite the Evil
Terrorists,-Dictators and Political Oppressions,-
Islamic everyday Living and Culture yet Upholds
and Practices the Uniqueness of Brotherhood.
Family and Neighbors still Thrive in Peaceand
Harmonious Accord. The Muslim Diametric of
Moral Varsity Still Resonates in the Sanctity of
Family and Communal Life. One Man, One
Woman,-nor Sexual "same sex" exists, regarded as
Perverse and Violating Universal Nature's Laws,
Hence No Aids or HIV Diseases exist,-as
Recorded by the United Nation's Physician
General.

We can Adopt Nature's Wisdom of Opposite
Plus and Minus Electrical Chemical Harmony,-to
Avoid untimely Tragic Diseases and Deaths. We
speak of. Extol Nature's "Eco System", when our
own Human Eco Immune system is Assailed,
assaulted and Diseased, Modern Man, is Nature's
Most "Endangered Species"!

Our Commercial Monopolies have Subjugated
and enslaved average American Humanity,-in
Fierce economic Inequality. "Free Enterprise",
now monopolized by American and Global Self-
Interest Greed and Avarice Stranglehold American
Majority,-Families Struggling to meet the Bare
Necessities of Life,-Food Stamps and Economic
assistance,-once ashamedly the last alternative,
presently Vitally Needed for impoverished

Majority ofAmericans.

America's Commercial Wealth once Pioneerd Fairness and Ethical Standards,-practised by Henry Ford, and Notably Andrew Carnegie. The "Common Good" was Heralded, especially by Carnegie, Who Solemnly believed that Wealth should Return to America's vast Populace, to be Utilized for Greater Prosperity and Good.

What a Fatal Contrast to Today's Economic Servitude and Horrific Inequality, Suffered by the American Masses!

Only Now,-do We Recognize Islam's Great Scientific Discoveries, 18 plus Centuries ago,- Which Ignited Medieval Dark Europe's Renaissance,-with Scientific, Mathematical Axioms, dispelling Christendom's Vampires and False Superstitions for Real Scientific Factual Truths. Yet to Most Westerners,-Remains the Astounding Facts and Reality, that All Islamic Science emerged from Religious Quranic Tenets and Revelations. Islam has Proven and Attested in Realistic Historical Factuality, that Religion and Science are not only Compatible, but necessarily One. God, the Supreme Creator, Unique Genius of all geniuses, -made possible to Man's Comprehensive Brain,-all knowledge possible from "Al Aqal". the "Infinte Universal Mind", "The First Thing that God Created for the Man, was Reason", -Al Quran, Islam can and will Reconcile Science and Religion.

As our Astronauts search Vast Space,-Light Years billions of miles Distant,-Universes and Solar Systems yet to be Discovered,-How Pray tell,-can he or she not Marvel that One Superest Brain and Power Has Indeed in Truth -in Reality Created and Shaped All Universal Nature!! We have ceased to Reason altogether!!! "Black Holes" were Discovered by Islamic Scientists Who Affirmed "created Life", and "destroyed" as well, Apparently, the Universal Will of God is Omitted,- as Originator of His Own Universe, as all our Science-Astronauts acclaimed Atheists, or Agnostics, have lost all Sane Rationality, Consumed, overwrought with their own Egos!

The Holy Prophet's First Command was that Muslims Search and Explore the Earth and Zodiac Heavens,-to Glean Understanding and Scientia, One Hundred Years post the Prophet's Life, the Muslim Geologists had measured the Circumference of the Earth. Electrons and Protons were discovered, that produced the world's first Generator, lighting the streets of Cordova, centuries before electricity was discovered in the West, Muslim Geologists discovering the Earth a Round Sphere, affirmed thusly, while Europe fancied the Earth flat,

The Muslim Mathematicians created the Decimal system, Numbers,-both Qualitative and Quantitative,-as the Spiritual significance was an integral part of analysis. It was Uniquely, Omar Khayyam's intricate Formulas that made possible Mankind's Flight to the Moon and Planetary travel, The Muslims' belief in One God as Infinite, not triune, established the Zero, "Ziffer", and the basis for all Mathematics, the Alphabet supplanting Roman numerals,

We Christians are Mesmerized, Overwrought by "Sin", and its foreboding Sinful Implications, Islam has no such Sin-Ridden Complex Re-Constituting All Mankind Derives from our Divine Nature and Virtuous Attributes from our Heavenly Father in Heaven, "The Kingdom of Heaven is Within you", is Reaffirmed by Islam. "God Reigns in the Heart of Man", Although Evil is prevalent,-it has no Real lasting efficacy, but must Perish as with all evil deeds, "GOOD or GOD is the Only Reality", Al Quran Reveals, Jesus Christ is indeed "El Messiah" The One True Messiah sent to All Humankind. To Obey all his Commandments, Above All the First Commandment "To Worship God, with All thy Mind, Soul and Heart", To Worship Jesus, instead of God First, Violates Jesus's own Commandment. God is not Flesh, or limit embodied. But God is Infinite, Above and Beyond all human devisement. All Glory and

Praises to God Creator.

Who Created Man from a Clot.-and Taught him by the Pen, What he Knoweth not"! "God neither Begets, nor is begotten and in His UNCREATED INFINITY,-All Everlasting Eternal Glory"!

Reveals Al Quran,

Albeit, we are "created in God's Image", (with some 150 Millenniums to go),-Islam Reveals that Man's Nature "Fitrah",-Originally derives from God's Nature. "Al Fitrah" will One Day Realize our Divine Nature and Being. "GOD Has Created Man in His Own Nature, "Al Fitrah",-Ultimately Humankind's Futurity and Destiny! Since our One Day Divine Fused Nature Evolves into Good or God Fruition and Perfection.....Man is Essentially "Sinless", with God's Inbred Potential for Perfection, Hence NO "original sin" exists in Islam. God Created Man out of LOVE and Infinite Beneficent Mercy. "Why Look Ye. O Man for Miracles,-when Ye are my Greatest Marvel and Miracle?" Speaks God to our Man Creation. "It is I. Who Shaped ye in your Mother's womb,-Gave ye Sight and Hearing, and Comprehension above all Earthly Creatures, "Man" translates "Innsaun" in Arabic,-"The Knowledgeable Creature". "I Have Created you, in My Stead, to Rule, over the Earth,-Entrusted you with Care of All Universal Nature".

To Express the Heart of Islam: "God Created Man in His Own very Nature and Being. AL Fitrah", Sura 30:30,

How be we. So Blest,-the Likes of thee and me!!

What be our Optimal Blessed Destiny,-"Kismet"!!

Hell,-Jahundum,-means "Distance from God/Good".... whereas Heaven,-is "Nearness", Jubilance and Peace" in God. The Devil, "Shatan" has No same "billing" in Islam, as in Christiandom. We create our own "devils" and "hell", in the Here and now!

God's Infinite Mercy is Never-ending. In the Hereafter, all Evil will be purged, and "New Rivers will Flow to purify and redeem all living beings".

One Mystic Muslim approached God, ascribing God's unfailing Mercy. He exclaimed to God,-"If they knew of your Boundless Mercy.-they wouldn't even bother to pray". "You keep your secret, and I'll Keep Mine" came the Reply, Multiple Muslim Mystics expressed their Ecstatic Prayerful Oneness with God,-centuries before Christian Saints. One Famous Woman, Mystic-Poet, Rabiya Al Adawiya Wrote... "O God, if I Worship you for Fear of Hell,-Burn me" and if I Love Thee for Hope of Heaven.-Withhold Heaven from me. But, If I Worship Thee for Thy Own

Sweet Love's Sake, Grant me Love's Undying Eternity in Thee"!

Ibn Arabi exclaimed "LOVE/GOD is The Source of All Love "Were it Not for GOD/LOVE, Love would not be"! One Orientalist Marveled at the Islamic God's Nearness and Liaison in Man, likening the Moon's Shimmering Reflection on the Waters below, to God's Glory Imaged in Man, Man, the Instrument and "Reason" for God's proximity. "God would not See His Own Glory, unless Reflected on the Waters below", Man's Simile of God's Imaging Light.

The Holy Prophet Revealed that All Prayer should Inspire Oneness in God. The Apex of all Prayer, Perfects in God Creator's Revelation,-"When My Faithful Believer Comes Nearest Me,-I become his Eyes through which he sees. his Ears, from which he hears, and Hands that grasps Mine"!

For the Fearful. Unduly Mesmerized by "brimstone and fire", -for Redemption and "Salvation".-The Heart of Islam Reveals that Man is "Saved", by Virtue of his Indwelling

God Nature, "Al Fitrah".

Chapter 18

EPILOGUE

WHAT IS TRUTH?

"Truth" Exacts No clear Meaning for most Individuals.

Truth is espied as Particular to One's personal circumstances, whatever they be. On Consensus,- in general, Truth stands as agreement on various Issues or Topics, albeit Politics, Religion or perceptions of Success,-beit Wealth, Health, Wellbeing, Creative aspirations, or Spiritual Evolvement, and Attainment of Inner Peace and Bliss.

The Profounder aspects of Truth, lie in One's True Heart's Concept of Perennial Permanence of Truth, versus Transient Ephemeral notions. Truth, as such, must infer a Timelessness. Truth, as such, must incorporate an Eternal factor, or State of being. Such a Timeless quality or State necessarily addresses and Attaches to God, or an Eternal Being, over and beyond our transient passing Mortal lives.

Only Islam Directs Man's Unique Oneness of Liaison with the Timeless Eternal Creator. "Hope of Salvation" is already revealed in Surah 30:30, -

that attests "Man is forged, Created in God's Own Divine Nature, "Al Fitnah". By Virtue of God's Eternal, Ever living Spirit, Substance and Nature, "Fitrah" in Man,-his Potential for Good/God is Certified. Unless he or she Violates Good/God's Immutable Laws.

God/Immutable Truth Lies in Man's Heart, in the Here and Now. The Supernatural is Natural, herein our grasp

and Embrace. All Prayer leads to Oneness and Loving Union in God,-ineffable Bliss and Peace, reveals Islam.

No "Original Sin" exists in Islam we, of Humanity are created out of Love/God. We are "Saved", by God's Innate, Indwelling Nature and Being. Reiteratively, "Al Fitnah".

The Good among Humankind, who live and Flourish in God/Good's Ordinances and Brethren Love, inherit Heaven Now, and those afflicted with Evil deeds, Suffer Hell and Misery Now, before the Hereafter. "Hell" is not a Place, but State of being,-raging Fire to Purify the Soul free from iniquity....God's Love, in reverse. "There are New Rivers of Redemption and Rebirth that Purify the soul...

My Mercy Exceeds My Wrath", reveals "The Infinite Compassionate, Forever Merciful

Creator"!

"All man-contrived Evil perishes, short-lived annihilation.

I, Good/God AM The Only Reality"!   Reveals Al Quran.

www.ingramcontent.com/pod-product-compliance
Lightning Source LLC
Chambersburg PA
CBHW060825170526
45158CB00001B/87